Data Science for Civil Engineering

This book explains the use of data science-based techniques for modeling and providing optimal solutions to complex problems in civil engineering. It discusses civil engineering problems like air, water and land pollution, climate crisis, transportation infrastructures, traffic and travel modes, mobility services, and so forth. Divided into two sections, the first one deals with the basics of data science and essential mathematics while the second section covers pertinent applications in structural and environmental engineering, construction management, and transportation.

Features:

- Details information on essential mathematics required to implement civil engineering applications using data science techniques.
- Discusses broad background of data science and its fundamentals.
- Focuses on structural engineering, transportation systems, water resource management, geomatics, and environmental engineering.
- Includes Python programming libraries to solve complex problems.
- Addresses various real-world applications of data science–based civil engineering use cases.

This book is aimed at senior undergraduate students in civil engineering and applied data science.

Data Science for Civil Engineering
A Beginner's Guide

Rakesh K. Jain
Prashant Shantaram Dhotre
Deepak Tatyasaheb Mane
Parikshit Narendra Mahalle

CRC Press
Taylor & Francis Group
Boca Raton London New York

CRC Press is an imprint of the
Taylor & Francis Group, an **informa** business

Designed cover image: © Shutterstock

First edition published 2023
by CRC Press
6000 Broken Sound Parkway NW, Suite 300, Boca Raton, FL 33487-2742

and by CRC Press
4 Park Square, Milton Park, Abingdon, Oxon, OX14 4RN

CRC Press is an imprint of Taylor & Francis Group, LLC

© 2023 Rakesh K. Jain, Prashant Shantaram Dhotre, Deepak Tatyasaheb Mane and
Parikshit Narendra Mahalle

ISBN: 978-1-032-32780-8 (hbk)
ISBN: 978-1-032-32781-5 (pbk)
ISBN: 978-1-003-31667-1 (ebk)

DOI: 10.1201/9781003316671

Typeset in Times
by MPS Limited, Dehradun

Contents

About the Authors

Rakesh K. Jain obtained his B.E. degree in civil engineering from R. D. University Jabalpur, Madhya Pradesh, India and M.E. degree in town and country planning from University of Pune, Maharashtra, India. He received his PhD in civil engineering specialization in wireless communication from Rajiv Gandhi Technical University, Bhopal (MP), India. Currently, he is working as a director of JSPM's Rajarshi Shahu College of Engineering, Tathawade, Pune (An Autonomous Institute, affiliated to Savitribai Phule Pune University (SPPU), Pune, India). He has more than 35 years of teaching and industry and research experience. He worked as a principal of Dr. D.Y. Patil Institute of Technology, Pune, India for almost 10 years. Prior to this designation, he had worked as a lecturer, assistant professor, and head of civil engineering at D. Y. Patil College of Engineering, Pune, India. Apart from being an academician, he worked in an industry as a site engineer with Matrix Engineers, Pune, India. He has been working as an expert member of the Civil Engineering Program in the Expert Committee for accreditation by NBA, New Delhi in the last five years. Presently, he has been promoted to work as a chairman for the same committee. He has been honored by All India Council of Technical Education, New Delhi, India by selecting him as a "Margdarshak'' (Mentor) for guiding other institutions for accreditation by NBA. He is functioning as an ambassador for mentoring five non-accredited institutes for accreditation by NAAC, under the "Paramarsha" (advisor) scheme sanctioned by UGC (University Grant Commission). He was honored as "Best Principal" by SPPU, Pune, India, "Best Teacher", and "Certificate of Excellence for Leadership" Dr. D. Y. Patil Vidya- Pratishthan Society, Pune. Also, he is an awardee of the "Visvesvaraya Construction Excellence Award" by Construction & Business Record, Mumbai, India. He was a member of the Board of Studies in Civil Engineering under Faculty of Engineering (FOE), Member Academic Council, Chairman Board of Studies in Civil Engineering under FOE, and Member Research & Recognition Committee and Member. He was chairman of BOS (Civil) for Railways (Adhoc) under the Faculty of Technology, SPPU, Pune. He was a member of the Standing Technical Advisory Committee (STAC) of PMC and participated actively in the development process of the Pune City. He is a member of the Confederation of Indian Industry (CII) and Mahratta Chamber of Commerce, Industries, Agriculture (MCCIA). He has published more than 75 research publications and 5 authored books to his credit by CRC Press. He has five patents to his credit. He is a recognized PhD guide of SSPU, Pune and six students have successfully defended their PhD. Presently, he is guiding five candidates for their PhD. He has received grants under Modernization and Removal of Obsolescence (MODROBS), Research Promotion Schemes (RPS), and Seminars and Workshops from AICTE, New Delhi and SPPU. He has visited a few countries like USA, Germany, Singapore, Malaysia, and Thailand.

Prashant Shantaram Dhotre has completed his B.E. degree in computer science and engineering from Shri Ramanand Teertha Marathwada University, Nanded, in 2004 and M.E. degree in information technology from Savitribai Phule Pune University, Pune, in 2010. He received a doctorate degree in computer science and engineering from Aalborg University, Denmark in 2017. His PhD thesis title was "Systematic Analysis and Visualization of Privacy Policies of Online Services". His PhD work has resulted in a high-level acceptance by the Aalborg University, Denmark and published the work in Danish News/ Media. Currently, he is working as a professor in the Department of Information Technology at MIT Art, Design and Technology (MIT ADT) University, Pune, India. He has more than 18 years of teaching and research experience. He has carried out funded research projects on "Location and Identity Privacy in Internet of Things" and "Drowsy Driver Detection using Image Processing" by Savitribai Phule Pune University, Pune. He has 40 publications in International Journals and Conferences. One of his research papers was awarded the "Best Paper" in an International Conference on Intelligent Computing and Communication ICICC – 2017 Springer Series. He has authored eight books like *Internet of Things Integrated Augmented Reality* (Springer Nature Press), *Context-aware Pervasive Systems and Application* (Springer Nature Press), *Theory of Computations, Data Structures*, and *Machine Learning* (Technical Publication). A special book chapter was published in a book titled *Cyber Security and Privacy: Bridging the Gap*, from River Publication in Europe. His two chapters have been published in a book titled *Handbook on ICT in Developing Countries: Next Generation ICT Technologies* from River Publication in Europe. He is a corporate trainer and has conducted training for employees of several industries like Cummins, Siemens, etc. He has been invited as a session chair for international conferences like ICINC, ICCUBEA, etc. He is working as SPoC for International Collaboration between Indian Institutes (Sinhgad, Dr. DY Patil Pimpri, MITADT, etc.) and CMI, Aalborg University, Copenhagen, Denmark. As an outcome of this collaboration, more than 70 Indian students have completed an Internship abroad and 2 candidates have successfully completed their postdoc from abroad. As a part of his research, he has delivered several talks on research methodology, how to write research papers, use of reference manager and citation tools, how to complete a PhD in time: a roadmap, etc. at several institutes in Maharashtra. Also, from an academic perspective, he has been invited to talk about academic topics like use of data structures in industry, Internet of Things, theory of computation, etc. He is a recognized PG guide of Savitribai Phule Pune University and guides several students in the area of machine learning, big data security information privacy, etc. His research interests include algorithms, privacy, Internet of Things, and turing machine. He has visited a few countries like Denmark, France, Sweden, Germany, Norway, China, and UAE.

Deepak Tatyasaheb Mane has completed his B.E. degree in information technology from Dr. Babasaheb Ambedkar Marathwada University, Aurangabad, in 2004 and M.E. degree in computer engineering from Bharti Vidyapeeth Deemded University, Pune, in 2011. He has received a doctorate degree in computer science and engineering from Swami Ramanand Marathwada University in 2019. His PhD thesis title was "Design and Development of Supervised Clustering Algorithms". Currently, he is working as an associate professor in the Department of Computer Engineering at Vishwakarma Institute of Technology, Pune, India. He has more than 15.8 years of teaching and research experience. He has carried out funded research projects on "Amalgam Based to design Translator and synthesizer for English to Sanskrit Language" by Savitribai Phule Pune University, Pune. He has published 20 research papers in reputed international journals like Springer, Elsevier, Inderscience, IGI Global, etc. He has authored four Techmax publication books like *Principles of Programming Languages, Fundamental of Programming Languages, Programming Paradigms*, and *Machine*. He is a reviewer for several journals as well. He has been invited as a session chair for international conferences like ICTIS 2020 Ahmadabad India, ICT4SD 2020, GOA, India, etc. As a part of his research, he has delivered several talks on deep learning', machine learning techniques, high-performance computing, machine and deep Learning, etc. at several Institutes in Maharashtra, India. He is a recognized PG guide of Savitribai Phule Pune University and is guiding several students in the area of pattern classification, data science, machine learning, deep learning, etc. In 2020, he received the I2OR national "Eminent Young Researcher Award" and in 2021, his research profile was selected for "BEST Researcher Award" in International Scientist Award 2021 on Engineering, Science & Medicine in Goa, India. His main research interest focuses on machine learning, pattern classification, supervised clustering, and neural network.

Parikshit Narendra Mahalle is a senior member IEEE and is professor and head of the Department of Artificial Intelligence and Data Science at Vishwakarma Institute of Information Technology, Pune, India. He received his PhD from Aalborg University, Denmark and continued as a postdoc researcher at CMI, Copenhagen, Denmark. He has 22+ years of teaching and research experience. He is a member of the Board of Studies in Computer Engineering, ex-chairman Information Technology, SPPU and various universities and autonomous colleges across India. He has 9 patents, 200+ research publications (Google Scholar citations: 2,150 plus, H index-22, and Scopus Citations are 1,100 plus with H index-15) and authored/edited 40+ books with CRC Press and other publishers. He is editor in chief for *IGI Global–International Journal of Rough Sets and Data Analysis*, associate editor for *IGI Global–International Journal of Synthetic Emotions, Interscience International Journal of Grid and Utility Computing*, and member of the editorial review board for *IGI Global–International Journal of Ambient Computing and Intelligence*. His research interests are machine learning, data science, algorithms, Internet of Things, and identity management and security. He is a recognized PhD guide of SSPU, Pune, guiding seven PhD students in the area of IoT and machine learning. Recently, five students have successfully defended their PhD. He is also the recipient of "Best Faculty Award" by Sinhgad Institutes and Cognizant Technologies Solutions. He has delivered 200 plus lectures at the national and international level. He is also the recipient of the best faculty award by Cognizant Technology Solutions.

Preface

"If you can't explain it simply, you don't understand it well enough."
- ALBERT EINSTEIN

With advancements in broadband technology and the semiconductor industry, the cost of the Internet and sensory devices is decreasing at a faster rate. The Internet is getting cheaper and faster day by day, resulting in a reduction in the cost of Internet connectivity. Due to the transformation, more devices are connected to the Internet and are causing data overflow i.e., big data. Fortunately, database management systems i.e., extracting, transforming, and loading the data are mature and the main challenge today is how to make sense of this big data. All information technology leaders and researchers are facing the main challenge of drawing meaningful insights from this big data and extending these insights for business intelligence. The main objective of this book is to guide beginners about the application of data science to the data sets in civil engineering. The text is divided into three parts: concepts and foundations, applications, and research areas.

Research and development play a major role in the economic growth and the GDP of a nation. This book fundamentally aims to provide the basic skills required for the data science role amongst the readers by enhancing the ability to understand, connect, apply, and use the data science to the problem in civil engineering. This book provides the beginners with an in-depth knowledge of different categories of the problems in civil engineering that require a data science approach to reach major outcomes like recommendations, forecasting, estimation, decision, and analysis. This book is majorly concerned with providing mathematical foundations of data science and the skills required in developing the expertise of identifying the category or class into which the problem at hand falls and eventually how to apply data science to it. This book is targeted at academicians, researchers, students, and professionals who belong to all civil engineering disciplines. The main disciplines of civil engineering covered in this book are geomatics, water resource engineering, and environmental engineering.

This book is aimed at providing timely and useful content to the mass readers, especially to researchers and PhD scholars to whom data science and its applications become inevitable, irrespective of the branch of study in civil engineering.

The main characteristics of this book are:

- An easy, crisp, and concise overview of data science
- Provides a strong foundation for the readers regarding the basics of data science and analytics, its relevance and needs, especially for beginners.
- Equips the readers with the ability to identify and categorize the problem at hand
- Provides an in-depth understanding of the various categories/classes of problems in civil engineering

This book is specifically designed for beginners who want to get acquainted with the emerging field of data science and its application in civil engineering data sets for better business intelligence. Focusing on the identification of the category of problem and the application of appropriate data science techniques are key highlights in this book. This book is useful for undergraduates, postgraduates, industry, researchers, teachers, and research scholars in engineering and we are sure that this book will be well received by all stakeholders.

Rakesh K. Jain
Prashant Shantaram Dhotre
Deepak Tatyasaheb Mane
Parikshit Narendra Mahalle

1 Introduction

Deepak Mane and Prashant Dhotre

1.1 INTRODUCTION

The most demanding job in the twenty-first century is data science (Kumar et al., 2019). Every company is looking for employees who have experience with data science. The field of data science has swept the globe. In today's cloud-based corporate world, everything moves quickly, collecting and compiling data at a breakneck pace. The options for managing and employing data operations are limitless, with new data science applications being discovered every day and minute, resulting in new job opportunities across industries and sectors, as well as new career routes. So, if you're thinking about a career in data science, you'll need to grasp what data science is.

In the current era of cloud-based enterprise, data science plays a significant role. There are numerous paths through which new applications of this sector are continually growing, whether it is data usage or managerial activities. Learning data science allows you to improve your decision-making process and render the pattern of discovery and predictive analysis even better.

The demand for data scientists is enormous and growing all the time. Today, practically every area of the economy relies on this position. The process of accumulating history and present data to forecast performance and prospects is a useful component of this domain.

1.1.1 PURPOSE OF DATA SCIENCE

The basic goal of data science is to find the proper patterns in the data that has been acquired or collected. To evaluate the data and gain insights from it, a variety of statistical methodologies are used. Data scientists must inspect their data throughout the process, from data extraction to backbiting and preprocessing. A data scientist's other responsibilities include developing predictions based on the data. Data scientists are tasked with drawing inferences from data. Data scientists can use these insights to assist firms in making more informed and efficient business decisions. This chapter will be divided into sections to help you better understand the function of the data scientist.

To the business, data is like new oil. We are living in the Industrial 4.0 era. The age of big data and artificial intelligence has arrived. There is a massive data explosion that is resulting in new technologies and more efficient products. A day generates around 2.5 exabytes of data (Kumar et al., 2019). Over the last decade,

DOI: 10.1201/9781003316671-1

there has been a huge increase in the demand for data. Many companies have made data the focal point of their business services.

The term "data science" is relatively recent. There were statisticians before data science. These statisticians have qualitative data analysis knowledge, and they are used by firms to analyze overall sales and performance.

The study of statistics has been fused with computer science, resulting in data science, as a result of the introduction of computing processes, data storage on the cloud, and the usage of analytical tools.

Early data analysis activities include using surveys and identifying solutions to broad problems. A survey of numerous youngsters in a region, for example, can lead to important judgments about school development in that area. As a result, with the help of computers and computing technologies, the decision-making process has grown easier and simpler. As a result, employing data science to solve statistically based complicated problems has become straightforward.

Businesses began to appreciate the usefulness of data as it became more widely available. Many items designed to improve the customer experience reflect its importance. The industry was looking for someone who could help them unlock the power of data. The information assists you in making the best business decisions and increasing your profitability. We've also given the organization the opportunity to study client behaviour based on purchasing patterns and take appropriate action. Companies were able to strengthen their revenue models and offer higher-quality items for their customers as a result of the data.

Data is essential for product creation, much as power is required to run home appliances. As a result, the development of a user product necessitates the collection of both qualitative and quantitative data. That is what propels the product and allows it to exist. The data scientist, like a sculptor, separates the data before creating anything useful from it. It's a difficult assignment, but data scientists need the necessary skills and expertise to get the job done.

Data is required by the industry in order to make educated judgments. Data science is the process of transforming raw and unprocessed data into useful information. As a result, the industry recognizes the importance of data science. A data scientist is a wizard who can utilize data to produce magical and meaningful results. Data scientists with experience know how to extract useful information from the data they encounter. They steer the company in the proper direction.

He/she is an expert in data-driven decision making, which the company requires. Data scientists are experts in a variety of computer science and statistics domains. They tackle corporate problems by applying their analytical skills.

Data scientists are used to solving problems and are entrusted with identifying meaningful patterns in data. The purpose is to find and compare duplicate samples in order to draw conclusions. Data science necessitates a collection of techniques for extracting information from unprocessed and massive data sets. The roles of data scientists will differ depending on the application. Data collection and storage, as well as the treatment of organized and unstructured data, are critical activities.

The role of data science is to focus on data analysis and management. The assignment is dependent on the company's area of expertise. As a result, data scientists must also be familiar with the organization and its many business areas.

Businesses, as previously said, rely on meaningful data to function. They require it in order to develop a data-driven decision-making model and improve customer service. The domains where data science can be used to make smart data-driven decisions are discussed in this section (I. Martinez, 2021).

- Better Marketing with Data Science: Data is used by the company to examine marketing efforts and generate the most effective advertisements. In many occasions, firms offer their items for stratospheric sums. This does not always produce the desired consequences. By gathering, researching, and interpreting customer feedback, businesses may generate better marketing. Companies actively analyze their clients' online behaviour to do so. Companies can also obtain a better understanding of the market by monitoring and reading client trends. As a result, businesses want data scientists to assist them in making educated decisions.
- Customer Acquisition Using Data Science: Data scientists assist companies in acquiring clients by evaluating and comprehending their requirements. This allows businesses to tailor their products to the wants of their customers. A company's understanding of its customers is based on data. As a result, data scientists' role is to assist businesses in identifying their customers and meeting their needs.
- Data Science in the Service of Innovation: Companies use a plethora of data to fuel innovation. By evaluating traditional designs and gaining insights, data scientists aid product innovation. Customers' ratings and feedback are analyzed, and the company assists companies in developing items that exactly match their ratings and feedback. Companies sort judgments and take the correct measures in the right direction based on customer feedback data.
- Using Data Science to Improve People's Lives: Customer data is crucial to making their lives better. To support clients' daily life, the healthcare business makes use of openly available data. Personal data and health history are analyzed by data scientists in this industry to produce products that address client needs. It's evident from the data-centric enterprise example above that each company uses data in a unique way. The way you use data is determined by your company's requirements. As a result, the data scientist's role is determined by the company's profitability.

1.1.2 BI AND DATA SCIENCE: WHAT'S THE DIFFERENCE

BI is used for business data analysis. Data source, necessary skills, focus, and other criteria can be used to discuss the distinctions between BI and data science. The data source in BI is structured. Data sources for DS include both organized and unstructured data. In BI, the analytical method is utilized for data analysis. In DS, the scientific method is employed to deal with the data in a more in-depth manner. Statistics and data visualization are two abilities required for BI. Machine learning, along with statistics and visualization, is a necessary ability for data science. The focus of BI is on current and historical data. Data science is concerned with not only past and present data but also forecasts for the future.

1.1.3 COMPONENTS OF DATA SCIENCE

Key components in data science are represented in Figure 1.1 and discussed below (Brodie, 2019):

- Statistics: This is the major aspect of data science that uses different methods of gathering, analyzing, and visualizing vast amounts of numerical data in order to derive useful conclusions.
- Domain Expertise: Domain expertise is the glue that holds data science self-possessed. Domain expertise refers to specific knowledge or abilities in a specific field. There are several domains in data science where domain specialists are required.
- Data engineering: It is a subset of data science that entails data acquisition, storage, retrieval, and transformation. Data about the data (metadata) is also added to the data throughout the process of data engineering.
- Visualization: The display of data in a visual context so that stakeholders can effortlessly capture what it means is known as data visualization. Data visualization allows you to quickly access the massive amounts of information included in your visuals.
- Advanced Computing: Advanced computing makes up the majority of data science. The source code of computer programs is designed, created, debugged, and maintained in advanced computing.
- Mathematics: Data science relies heavily on mathematics. Quantity, structure, space, and change are all studied in mathematics. A thorough understanding of mathematics is required of a data scientist.

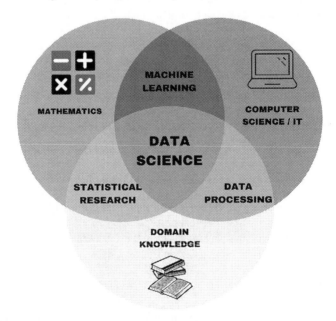

FIGURE 1.1   Components of Data Science.

- Machine learning: It is the backbone and important part of data science. Machine learning is the process of training a machine to mimic the functions of a human brain. To answer challenges and issues in data science, a variety of machine learning algorithms can be used.

1.2 DATA SCIENCE: AN OVERVIEW

Data science is a field of study that involves the use of a variety of processes, scientific methods, and algorithms to extract ideas from vast amounts of data. It aids in the discovery of hidden patterns in raw data. The evolution of mathematical statistics, large data, and its meaningful analysis has given rise to the data science field. Data science is a multidisciplinary field that allows you to extract knowledge from both structured and unstructured data. The use of data science is to convert a business/real problem into a scientific attempt, and then provide a real solution. Data science comprises scientific methods to study huge amounts of data (structured and unstructured) and that entails extracting relevant insights.

In a nutshell, data science is concerned with the following parameters (I. Martinez, 2021) and it is represented in Figure 1.2:

- Properly posing questions and assessing raw data.
- Using a variety of complex and efficient algorithms to model the data.
- Using visualization to gain a better understanding of the data.
- Analyzing data to find the end result and make decisions.

Let's say we want to travel by car from station A to station B. Now we need to make some decisions about which route will be the fastest to get to the destination, which route will have the least amount of traffic, and which route would be the most cost-effective. All of these decision elements will serve as input data, and we will

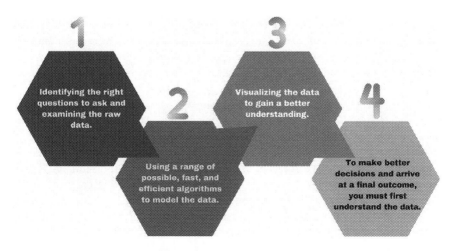

FIGURE 1.2 What Is Data Science?

obtain an acceptable response from these decisions; therefore, this data analysis, which is a part of data science, is called data analysis.

1.2.1 NEED FOR DATA SCIENCE

Data was less and primarily available in an organized form a few years ago, which could be easily stored in Excel sheets and processed using business intelligence tools. However, in today's world, data is becoming so large that 2.5 quintals bytes of data are generated every day, resulting in a data explosion. According to studies, by 2020, a single person on Earth would create 1.7 MB of data every second (Kumar et al., 2019). To develop, grow, and enhance their enterprises, every company needs data. Every organization now faces a difficult problem in dealing with such a large amount of data. So we needed some complex, powerful, and efficient algorithms and technology to handle, process, and analyze this, and that technology came to be known as data science. The following are some of the most important reasons to use data science technology (Mikalef & Krogstie, 2019):

- Using data science technologies, we can turn enormous amounts of unstructured and raw data into actionable insights.
- Various companies, whether huge brands or startups, are using data science technology. Google, Amazon, Netflix, and other companies that deal with large amounts of data use data science algorithms to improve the customer experience.
- Data science is assisting in the automation of transportation, such as the development of a self-driving automobile, which is the transportation mode of the future.
- Data science can aid in a variety of forecasts, including polls, elections, and flight ticket confirmation, among others.

So, now that you know what data science is, you must investigate why it is vital. As a result, data has become the industry's fuel. It's the next generation of electricity. Data is required for businesses to run, grow, and improve. Data scientists work with data in order to help businesses make better decisions. Companies use a data-driven approach to obtain useful insights with the help of data scientists who analyze massive amounts of data. These insights will be beneficial to businesses looking to assess themselves and their market performance. Healthcare industries, in addition to commercial businesses, utilize data science. The technology is in high demand to recognize microscopic cancers and abnormalities at an early stage of diagnosis. Since 2012, the number of data scientist jobs has increased by 650%. According to the U.S. Bureau of Labor Statistics, 11.5 million jobs will be generated by 2026. On LinkedIn, the position of data scientist is also included as one of the top developing jobs. All of the data points to rising demand for data scientists.

1.2.2 ROLE OF A DATA SCIENTIST

It is important to know the data scientist and their responsibilities in various domains. The data scientists are responsible to handle both unstructured and structured data. The unstructured data gets a meaningful structure in an unprocessed raw format that necessitates considerable data pre-processing, cleaning, and categorization before a data set can be given a meaningful structure (Jiménez et al., 2018). The data scientist then studies and thoroughly analyzes this organized data in order to obtain knowledge from it, employing a variety of statistical approaches. In order to analyze, label, visualize, and get meaningful information, different methods are used statistically. The data scientist then estimates the occurrence of events and makes data-driven judgments using powerful machine learning techniques.

A data scientist uses a wide range of tools and techniques to spot duplicate patterns in data. Python, Hadoop, SQL, and R, and are just a few of the tools available. Data scientists are typically hired as consultants, where the role is to help with different processes and development of strategy. To put it another way, data scientists help firms make better business decisions by extracting relevant insights from data. Companies such as Google, Netflix, and Amazon, for example, are utilizing data science to generate erudite recommendation systems for their customers/users. Predictive analytics and forecasting approaches are also being used by a number of financial firms to forecast stock prices. Data science has aided in the development of smarter systems capable of making autonomous judgments based on past data.

It has manifested itself to construct a larger picture of artificial intelligence through its integration with developing technologies such as computer vision, natural language processing, and reinforcement learning (Jiménez et al., 2018).

1.2.3 PROBLEMS AND SOLUTIONS USING DATA SCIENCE

When it comes to using data science to solve a real-world problem, the first step is to clean and preprocess the data. A data set may be sent to a data scientist in an unstructured format with different irregularities. It's easier to examine and develop conclusions when the data is organized and erroneous information is removed. This procedure entails removing superfluous data, transforming data into a prescribed format, and dealing with missing numbers, among other things (Chen et al., 2021). A data scientist uses numerous statistical processes to analyze the data. Two sorts of methods employed in particular are descriptive and inferential statistics.

Let's consider a cell phone manufacturing company and you are a data scientist. You, being a data scientist, must study customers using your company's mobile phones. To do so, you must first examine the data thoroughly and comprehend the numerous trends and patterns involved. Finally, you'll synthesize the information and present it as a graph or chart. As a result, you solve the problem using descriptive statistics.

The data will subsequently be used to draw 'inferences' or conclusions. The following example will help us grasp inferential statistics: Assume you're looking for a lot of flaws that occurred during the production process. Individual testing of mobile phones, on the other hand, can take a long time.

As a result, you'll look at a sample of the phones available and draw a generalization about the number of defective phones in the overall sample. You must now anticipate sales of mobile phones for the following two years. Regression algorithms will be employed as a result. You'll use regression algorithms to forecast sales over time based on previous sales data.

Based on their annual wage, age, gender, and credit score, you'll also want to assess if buyers are inclined to acquire the goods. You'll utilize past data to see if clients are likely to buy (1) or (0). As there are two outputs classes,' you'll utilize a binary classification algorithm. We also utilize the multivariate classification algorithm to solve the problem if there are more than two output classifications. Supervised learning encompasses both of the aforementioned issues. There are also instances of data that haven't been classified. There is no output division into fixed classes as in the previous example. Assume you need to identify groups of possible clients and leads depending on their socioeconomic status. You'll use the clustering algorithm to find clusters or sets of potential clients because your previous data doesn't have a fixed set of categories. Clustering is a form of unsupervised learning method (Xue & Zhu, 2009).

Self-driving cars are becoming increasingly popular. Autonomy, or the ability to make judgments without human interference, is the guiding principle of self-driving cars. In order to deliver results, traditional computers required human input. The problem of human dependence has been handled by reinforcement learning. Reinforcement learning is the process of performing specified activities in order to maximize reward. Consider the following example to see what I mean: Assume you're teaching your dog to fetch a ball. The dog is then rewarded with a treat or a toy each time it retrieves the ball. If it doesn't fetch the ball, you don't give it a treat. If the dog returns the ball, it will be aware of the treat incentive. The same approach is used in reinforcement learning. We offer the agent a reward depending on its actions, and it will aim to maximize it.

1.3 BENEFITS, CHALLENGES, AND APPLICATIONS OF DATA SCIENCE

Data science is the scientific study of data. In order to gain insights, it is the process of extracting, analyzing, displaying, managing, and storing data. These data-driven insights help firms make better decisions.

In data science, both unstructured and structured data must be employed. It's a multidisciplinary field with roots in math, statistics, and computer science. One of the most sought-after jobs is data science, which has a large number of open opportunities and a strong pay range. So, that was a fast overview of data science; now let's look at its advantages and disadvantages. The discipline of data science is vast and has its own set of benefits and drawbacks. So, we'll weigh the benefits and drawbacks of data science here. This post will assist you in evaluating yourself and selecting the appropriate data science course.

1.3.1 Benefits of Data Science

The following are some of the numerous advantages of data science (Analytics Insight, 2021):

- It's in high demand: The field of data science is extremely popular. Job seekers have a number of possibilities at their disposal. On Linkedin, it's the fastest-growing career, with 11.5 million new jobs predicted by 2026. For a reason, data science is a field with a lot of job opportunities.
- There are a lot of job openings: Only a few people have all of the skills required to be an overflowing data scientist. As a consequence, DS has a lower saturation rate than other IT fields. As a response, DS is a broad field with many possibilities. The field of data science is on rising; however, there are a limited number of data scientists available.
- A High-Paying Job: A data scientist is one of the top ten careers. According to Glassdoor, data scientists make an average of nearly $116,100 a year. As an outcome, data science is a highly profitable career path.
- Data science is adaptable: Data science can be used in a variety of ways. It is frequently used in mental well-being, finance, consulting, and e-commerce businesses. Data science is a broad field with numerous applications. As a result, you'll be eligible to function in a wide range of industries.
- Data Science Improves Data: Businesses require data scientists to process and assess their data. They not only examine the data but also improve its quality. As a result, data science focuses on enhancing data and making it much more helpful to businesses.
- Data Scientists are Highly Sought After: With the help of data scientists, businesses can make effective decisions. Companies depend on information scientists and use their expertise to provide good outcomes to their clients. This promotes data scientists to a significant position inside the company.
- There Will Be No More Boring Tasks: Benefits of Data Science: Machine learning has aided the industry in developing better goods that are precisely designed for customer experiences. Recommendation systems, for example, are used by e-commerce companies to provide users with personalized insights based on past purchases. Computers are now capable of understanding human behaviour and making data-driven decisions as a result of this advancement.
- Data Science Has the Potential to Save Lives: The healthcare industry has greatly improved as a result of data science. Because of developments in machine learning, it is now easier to diagnose early-stage cancers. Many other wellness companies are also utilizing data science to help their clients.
- Data Science Can Help You Grow As A Person: Data science would not only offer you a lucrative career, but it will also help you develop personally. You'll be able to tackle difficulties with a solution-oriented perspective. Because so many data science jobs combine IT and management, you'll be able to get the best of both worlds.

1.3.2 CHALLENGES OF DATA SCIENCE

While data science is a profitable professional path, it also comes with its own set of problems. We must understand the limitations of data science in order to get the complete picture of data science. The following are a few of them (Kumar et al., 2019):

- The term 'data science' is a misnomer: Data science is a fairly broad word that lacks a precise definition. While it has become a term, defining the specific meaning of a data scientist is difficult. The specific role of a data scientist is determined by the field in which the organization specializes. While some have referred to data science as the fourth paradigm of science, opponents have dismissed it as nothing more than a rebranding of statistics.
- It's nearly impossible to master data science: Mathematics, computer programming, and math and science are all incorporated into data science. It is difficult to be an expert in all disciplines and have equal knowledge in each. Despite the fact that various online courses have sought to cover the skill gap that the data science sector is facing, due to the complexity of the discipline, becoming an expert in it is still unattainable. A person with a statistical background might not have been able to obtain computer science quickly enough to become a good data scientist. As a result, it's a fast-paced, ever-changing field that demands a constant grasp of the many parts of data science.
- A significant amount of domain knowledge is required: Data science also has the drawback of being reliant on domain knowledge. Without prior knowledge of statistics and computer science, a person with a significant background in these fields will find it challenging to solve data science problems. The same can be said for the other way around. A healthcare industry working on genomic sequence analysis, for example, will require a suitable employee with some genetics and molecular biology understanding. This enables the data scientists to make informed judgments that will benefit the firm. A data scientist from a different background, on the other hand, may find it challenging to gain specific domain knowledge. This also makes transitioning from one industry to another challenging.
- Random data can lead to surprising results: A data scientist examines the data and develops careful forecasts in order to aid in decision making. The data provided is frequently arbitrary and does not produce the intended outcomes. This can also fail to owe to inadequate management and resource allocation.
- The issue of data privacy: Data is the lifeblood of many industries. Data scientists assist businesses in making data-driven decisions. However, the data used in the process may infringe on customers' privacy. Client personal data is available to the parent company, which might lead to data leaks if security is breached. Many sectors have been concerned about the ethical challenges surrounding data privacy and its use.

1.3.3   Data Science Life Cycle

The data science life cycle is depicted in Figure 1.3.

The following are the major stages of the data science life cycle (Fox, 2014):

- Discovery: The first step is discovery, which involves asking the right questions. You must first identify the project's fundamental requirements, priority, and budget before starting any data science project. Before outlining the business issue on the first hypothesis level, we must first define all project demands, such as the number of people, equipment, time frame, data, and an end goal.
- Data Preparation: Inconsistencies in data, such as missing values, blank columns, and erroneous data format must be cleansed. Before you can model, you must first process, investigate, and condition the data. Your predictions will be better if your data is clean.
- Model Planning: Data can contain a variety of abnormalities, such as missing values, blank columns, and improper data formats, all of which must be cleaned. Before you can model, you must first process, investigate, and condition your data. The more accurate your forecasts are, the cleaner your data is.

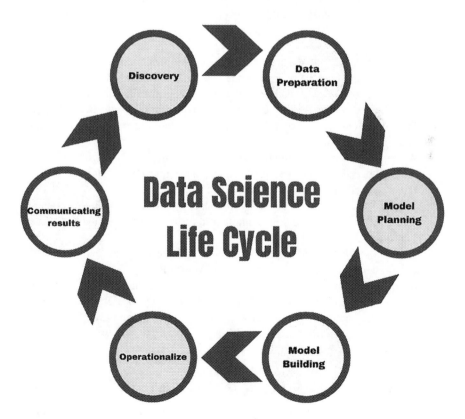

FIGURE 1.3   Life Cycle of Data Science.

- Model-Building: The model-building process begins in this phase. For the sake of training and testing, we shall construct data sets. To construct the model, we will use various strategies such as association, classification, and clustering. SAS Enterprise Miner, WEKA, SPCS Modeler, MATLAB, and other popular model building tools are just a few examples.
- Operationalize: During this phase, we will provide the project's summary report, as well as summaries, code, and technical documents. This phase includes a clear image of the complete project achievement and other components on a small scale before the major deployment.
- Communicate Results: We will give the project's summary sheet, and also abstracts, code, and related documentation during this phase. Before the big deployment, this phase offers a clear image of the whole project's achievement and additional components on a modest scale.

1.3.4 DATA SCIENCE AND ITS APPLICATIONS

Medical, finance, manufacturing, and transportation are just a few of the industries where data science has built a significant presence. It does have a wide variety of functions and applications. Health, e-commerce, research in manufacturing, multilingual agents, and transportation are just a few examples of data science applications (Kanal et al., 2017).

- Data science has played a crucial role in the healthcare industry. Using classification algorithms, doctors may employ machine vision software to make a diagnosis and see tumors at an early stage. In the genetics industry, data science is utilized to study and categorize genomic sequence patterns. Patients can also use virtual assistants to help them recover from physical and mental disorders.
- Amazon has a recommendation system that shows customers items based on past purchases. Data scientists have used machine learning to build recommendation systems that anticipate user preferences.
- Industrial robots have seized boring and repetitive work in the manufacturing facility. These self-contained industrial robots use methods like reinforcement learning and computer vision.
- Amazon's Alexa and Apple's Siri both employ speech recognition to comprehend people. Data scientists developed this speech recognition technology, which converts human speech into textual data. It also uses a number of machine learning to categorize and reply to user queries.
- Self-driving automobiles use autonomous agents that apply reinforcement learning and identification. Self-driving cars are no longer imagination, thanks to breakthroughs in data science.

1.4 DATA SCIENCE AND CIVIL ENGINEERING: OPPORTUNITIES

Data is consuming practically every industry, including some that you might not immediately associate with data, like construction management and civil engineering, for example. To take advantage of the massive volumes of data produced and pre-served by today's strong computing systems, big data technology is employed in construction management and other related industries. Information comes from a multitude of places, including computers, humans, sensors, and other data-gathering devices. Every structure constructed in the construction sector is backed by plans containing massive amounts of data.

Today's data, on the other hand, is supplemented by a large amount of data from sources such as building engineering logs, building staff, cranes, excavators, and materials logistics. CAD drawings, staff information, expenditures, bills, and project documents were all easily recorded using standard construction software in the past. Unstructured data, such as free-form text (mail and word docs), printed documents, and analog sensor data, is required by today's construction companies and civil engineers. Civil engineers can employ big data technologies to create the use of unstructured text that is hard to collect and evaluate manually. In fact, data is meaningless unless it could be reliably checked and examined. Engineers can now use metrics from a variety of sources to determine how and where a structure should be built, including architectural design concepts, environmental parameters, social site feedback, and stakeholder objectives. Engineers can also use prior construction data to predict problems and avoid project delays. The five ways that big data helps civil engineers achieve project goals are highlighted in the following entries (Goulet, 2020).

- A Better Understanding of Future Initiatives: Engineers can use big data technology to construct enormous infrastructures while avoiding difficul-ties that would otherwise arise. Public transit, for example, is an essential component of any modern city's infrastructure. Civil engineers use big data solutions to figure out how to improve public transportation's en-vironmental effect by lowering the emission of fossil fuels and con-sumption of energy. Civil engineers and firms like Black & Veatch may use the technology to make the most efficient use of land patterns and design smart cities that use the most economic, practical, and ecologically friendly transit systems available. Engineers also use data to develop marketing strategies in order to gain a better grasp of how to grow their firm and attract new clients.
- A more thorough examination of previous initiatives and large construc-tion projects can cost a lot of money. For example, operating costs at Hong Kong International Airport (HKIA) were $25.2 billion in 2016 and $34 billion at Dubai's Al Maktoum International Airport in 2014. Civil en-gineers' expertise is required for the efficient design, building, and ongoing management of such massive buildings. For large-scale projects like the Dubai and Hong Kong airports, stakeholders will usually recruit engineers having two or more generations of expertise. Big data technology is now used by civil engineers to analyse data from a number of sources and determine the right plan of action for proper management.
- Construction Project Planning, Simulation, and Tracking: Construction projects consume a huge amount of resources and produce a lot of data.

Big data is a natural fit for the construction industry, which has long been associated with extensive computer and accounting. Notwithstanding this link, the construction sector has lagged behind other industries in adopting big data technology. Construction companies, on the other hand, have only recently begun to analyse massive unstructured data sets using real-time, cloud-based tools. The strategy has the potential to improve collaboration between designers, technologists, and property owners.

- Transport Network Modern Analysis: Transportation researchers are using big data technologies to study and analyze massive amounts of data gathered by Iowa traffic infrastructure. Experts hope to enhance traffic management and detect incidents more quickly. The Realtime Analytics of Transportation Data Lab (REACTOR), for example, is based at the Center for Transportation Research and Education (CTRE), where scientists are developing a big data system that will identify and respond to traffic accidents, as its name suggests. The researchers are collaborating with the Iowa Department of Transportation (DOT) to enhance planning and organizational decisions. The system collects speed monitoring, current traffic data, video footage, weather patterns, and other variables statewide.
- Data is combined with artificial intelligence (AI) to generate a unique knowledge base.

1.5 SUMMARY OF THE BOOK

Analysts believe that artificial intelligence will bring in the next industrialization (AI). This is especially true in the construction sector, where civil engineers are tasked with shortening project delivery times, reducing environmental impact, and lowering costs. Artificial intelligence can manage project managers and engineers in achieving these objectives. Google, Oracle, and IBM, for example, store gigabytes of data in the cloud, which is then utilized by the technology. When big data technology creates solutions that humans can't perceive, AI will step in to ensure that skilled civil engineers don't miss out on an opportunity to improve. Big data analytics can give financial data, accurate event notifications, and complex analysis in real time. Civil engineers may use technology to quickly analyze and assess data, avoiding project errors that can mean the difference between winning and losing. As a result, the next generation of talented professional engineers will have access to data that was previously unavailable due to traditional construction procedures.

As data science has become more diverse and applied, the library of techniques has grown and expanded. The purpose of this book is to provide readers with a broad understanding of data science and the methodologies that may be utilized to address civil engineering challenges. The principles of both theoretical and practical elements are addressed in the disciplines of water resources/hydrological modelling, geotechnical engineering, construction engineering and management, and coastal/marine engineering. Drought forecasting, river flow forecasting, evaporation modeling, dew point temperature estimation, concrete modeling compressive strength, groundwater table forecasting, and wave heights height forecasting are only a few of the complicated civil engineering topics addressed.

- With regard to civil engineering, we have exclusive information on machine learning and data analytics applications.
- Many machine learning techniques are used in a variety of civil engineering specialities.
- Gives examples of how and where machine learning techniques might be used to solve problems.
- Water resources and hydrological modeling, geotechnical engineering, construction engineering and management, coastal and marine engineering, and geographic information systems are all topics covered in this course.

REFERENCES

Analytics Insight. (2021). *7 Benefits of Data Science that can Benefit your business.* DATA SCIENCE LATEST NEWS. https://www.analyticsinsight.net/7-benefits-of-data-science-that-can-benefit-your-business/

Brodie, M. L. (2019). Applied Data Science. *Applied Data Science, June.* 10.1007/978-3-030-11821-1

Chen, Y., Leung, C. K., Li, H., Shang, S., Wang, W., & Zheng, Z. (2021). A data science solution for supporting social and economic analysis. *Proceedings – 2021 IEEE 45th Annual Computers, Software, and Applications Conference, COMPSAC 2021,* 1689–1694. 10.1109/COMPSAC51774.2021.00252

Fox, P. (2014). *Data Life Cycle: Introduction, Definitions and Considerations.* 20. https://www.eudat.eu/sites/default/files/PeterFox.pdf

Goulet, J.-A. (2020). *Probabilistic Machine Learning for Civil Engineers.* MIT Press. https://mitpress.mit.edu/books/probabilistic-machine-learning-civil-engineers

I. Martinez, E. V. and I G. O. (2021). A survey study of success factors in data science projects. *2021 IEEE International Conference on Big Data (Big Data),* 2313–2318. 10.1109/BigData52589.2021.9671588

Jiménez, L., Solís, A., Science, C., Rica, T. D. C., Clara, S., Carlos, S., Rica, C., Science, C., Rica, T. D. C., Clara, S., Carlos, S., & Rica, C. (2018). *Different Roles in Data Science Field.* https://www.academia.edu/36246836/Different_roles_in_data_science_field

Kanal, E., Manager, T., Decapria, D., & Scientist, D. (2017). *Data Science Tutorial About us Eliezer Kanal Recent Projects: Cyber Risk Situational.*Data Science Tutorial August 10, 2017, Carnegie Mellon University https://resources.sei.cmu.edu/asset_files/Presentation/2017_017_001_503388.pdf

Kumar, S., Dhanda, N., & Pandey, A. (2019). Data science - Cosmic infoset mining, modeling and visualization. *2018 International Conference on Computational and Characterization Techniques in Engineering and Sciences, CCTES 2018,* 1–4. 10.1109/CCTES.2018.8674138

Mikalef, P., & Krogstie, J. (2019). Investigating the data science skill gap: An empirical analysis. *IEEE Global Engineering Education Conference, EDUCON, April-2019,* 1275–1284. 10.1109/EDUCON.2019.8725066

Xue, M., & Zhu, C. (2009). A study and application on machine learning of artificial intellligence. *IJCAI International Joint Conference on Artificial Intelligence,* 272–274. 10.1109/JCAI.2009.55

2 Mathematical Foundation for Data Science

Radhika Menon, Neha Shaikh, and Ambika Biradar

2.1 LINEAR ALGEBRA

Data science has applications in almost all branches of civil engineering. A strong mathematical theory forms the foundation of all data science models. Mathematical concepts like linear algebra, gradient-based optimization techniques and regression techniques are foundations of data analysis models. Therefore, in order to extract knowledge and hidden patterns from the raw data, it is essential to have understanding of all such core mathematical concepts. These mathematical concepts, together with some illustrations are covered in Sections 1 and 2.

Data, which may be an observation, feature or an attribute, are represented using vectors. These feature vectors or attributes form the input for classification and many of them share common properties related to specific data set. These vectors with similar properties form a vector space. To forms the basis for feature space; these feature vectors need to be represented as a linear combination of set of linear independent vectors or orthogonal vectors. Linear transformations are used in data analysis for projecting data from one space to another space with different dimension.

The first part deals with vector spaces, subspaces and affine subspace, the linear dependency and independency of vectors, basis and dimensions of vector space, linear transformation, eigenvalue and eigenvector of a linear transformation, and orthogonalization. Matrix decomposition methods such as LU decomposition, Cholesky's decomposition, and QR decomposition are discussed. This will help to reduce computational complexity involving matrix operations related to data analysis algorithms.

2.1.1 Vector spaces
2.1.2 Subspaces and norm
2.1.3 Linearly dependent and independent vectors
2.1.4 Basis and dimensions of vectors spaces
2.1.5 Linear transformation, Rank–Nullity theorem
2.1.6 Eigenvalue and eigenvector of linear transformation
2.1.7 Matrix factorization: LU decomposition, Cholesky's decomposition, and QR decomposition

DOI: 10.1201/9781003316671-2

2.1.1 VECTOR SPACE

A vector space V over a field K is a nonempty set V of objects, called vectors, on which are defined two binary operations called addition and multiplication by scalars from field K, subject to the axioms below. The axioms must hold for all vectors u, v, and w in V and for all scalars a and b in K.

    i. For every, $u, v \in V$ $u + v \in V$
    ii. For every, $u, v \in V$ $u + v = v + u$, (Commutative property)
    iii. For every $u, v, w \in V$, $(u + v) + w = u + (v + w)$, (associative property)
    iv. There is a vector (called the zero vector) 0 in V such that $u + 0 = 0 + u = u$ for each $u \in V$ (existence of additive identity)
    v. For each $u \in V$, there is vector $-u$ in V satisfying $u + (-u) = (-u) + u = 0$ (existence of additive inverse)
    vi. For each $u \in V$ and $a \in K$, $au \in V$
    vii. For each $u \in V$ and $a, b \in K$, $(a + b)u = au + bu$
    viii. For each $u, v \in V$ and $a \in K$, $a(u + v) = au + av$
    ix. For each $u \in V$ and $a, b \in K$, $(a \cdot b) \cdot u = a \cdot (b \cdot u)$
    x. For each $u \in V$ and $1 \in K$, $1 \cdot u = u$ (existence of multiplicative identity) [1]

Example 1: Verify whether the set $V = \left\{ M = \begin{bmatrix} a & b \\ c & d \end{bmatrix}, a, b, c, d \in R \right\}$ of matrices

of order 2×2 forms a vector spaces over a field $K = R$ set of real numbers under usual operation of matrix addition and scalar multiplication.

Solution:

i. Consider $M_1 = \begin{bmatrix} a & b \\ c & d \end{bmatrix}$, $M_2 = \begin{bmatrix} x & y \\ z & t \end{bmatrix}$ $a, b, c, d, x, y, z, t \in R$ in V then

$$M_1 + M_2 = \begin{bmatrix} a & b \\ c & d \end{bmatrix} + \begin{bmatrix} x & y \\ z & t \end{bmatrix} = \begin{bmatrix} a + x & b + y \\ c + z & d + t \end{bmatrix} \text{belongs to V since}$$

$a + x, b + y, c + z, d + t, \in R$

ii. Consider $M_1 = \begin{bmatrix} a & b \\ c & d \end{bmatrix}$, $M_2 = \begin{bmatrix} x & y \\ z & t \end{bmatrix}$, in V then

$$M_1 + M_2 = \begin{bmatrix} a & b \\ c & d \end{bmatrix} + \begin{bmatrix} x & y \\ z & t \end{bmatrix} = \begin{bmatrix} a + x & b + y \\ c + z & d + t \end{bmatrix} = \begin{bmatrix} x + a & y + b \\ z + c & t + d \end{bmatrix}$$

$$= \begin{bmatrix} x & y \\ z & t \end{bmatrix} + \begin{bmatrix} a & b \\ c & d \end{bmatrix} = M_2 + M_1$$

Thus, $M_1 + M_2 = M_2 + M_1$ for every M_1, M_2 in V

iii. Consider $M_1 = \begin{bmatrix} a & b \\ c & d \end{bmatrix}$, $M_2 = \begin{bmatrix} p & q \\ r & s \end{bmatrix}$ and $M_3 = \begin{bmatrix} w & x \\ y & z \end{bmatrix}$ in V then

$$M_1 + (M_2 + M_3) = \begin{bmatrix} a & b \\ c & d \end{bmatrix} + \left\{ \begin{bmatrix} p & q \\ r & s \end{bmatrix} + \begin{bmatrix} w & x \\ y & z \end{bmatrix} \right\} = \begin{bmatrix} a & b \\ c & d \end{bmatrix} + \begin{bmatrix} p+w & q+x \\ r+y & s+z \end{bmatrix}$$

$$= \begin{bmatrix} a+p+w & b+q+x \\ c+r+y & d+s+z \end{bmatrix}$$

$$(M_1 + M_2) + M_3 = \left\{ \begin{bmatrix} a & b \\ c & d \end{bmatrix} + \begin{bmatrix} p & q \\ r & s \end{bmatrix} \right\} + \begin{bmatrix} w & x \\ y & z \end{bmatrix} = \begin{bmatrix} a+p & b+q \\ c+r & d+s \end{bmatrix} + \begin{bmatrix} w & x \\ y & z \end{bmatrix}$$

$$= \begin{bmatrix} a+p+w & b+q+x \\ c+r+y & d+s+z \end{bmatrix}$$

Thus, $M_1 + (M_2 + M_3) = (M_1 + M_2) + M_3$ for every $M_1, M_2\ M_3$ in V

iv. for every M_1 in V there is a zero vector $O_{2\times2} = \begin{bmatrix} 0 & 0 \\ 0 & 0 \end{bmatrix}$ in V such that

$$M_1 + O_{2\times2} = \begin{bmatrix} a & b \\ c & d \end{bmatrix} + \begin{bmatrix} 0 & 0 \\ 0 & 0 \end{bmatrix} = \begin{bmatrix} a+0 & b+0 \\ c+0 & d+0 \end{bmatrix} = \begin{bmatrix} a & b \\ c & d \end{bmatrix} = M_1$$

v. for every M in V there is a unique vector $M' = \begin{bmatrix} -a & -b \\ -c & -d \end{bmatrix}$ in V such that

$$M + M' = \begin{bmatrix} a & b \\ c & d \end{bmatrix} + \begin{bmatrix} -a & -b \\ -c & -d \end{bmatrix} = \begin{bmatrix} a+(-a) & b+(-b) \\ c+(-c) & d+(-d) \end{bmatrix} = \begin{bmatrix} 0 & 0 \\ 0 & 0 \end{bmatrix}$$

vi. For each $M_1 \in V$ and $a \in R$, $aM_1 \in V$, $aM_1 = a\begin{bmatrix} x & y \\ z & t \end{bmatrix} = \begin{bmatrix} ax & ay \\ az & at \end{bmatrix} \in V$

vii. For each $M_1 \in V$ and $a, b \in R$, then

$$(a+b)M = (a+b)\begin{bmatrix} x & y \\ z & t \end{bmatrix} = \begin{bmatrix} (a+b)x & (a+b)y \\ (a+b)z & (a+b)t \end{bmatrix} = \begin{bmatrix} (ax+bx) & (ay+by) \\ (az+bz) & (at+bt) \end{bmatrix}$$

$$= \begin{bmatrix} (ax) & (ay) \\ (az) & (at) \end{bmatrix} + \begin{bmatrix} (bx) & (by) \\ (bz) & (bt) \end{bmatrix} = aM + bM$$

viii. For each $M_1, M_2 \in V$ and $a \in R$, then

$$
\begin{aligned}
a(M_1 + M_2) &= a\left(\begin{bmatrix} x & y \\ z & t \end{bmatrix} + \begin{bmatrix} p & q \\ r & s \end{bmatrix}\right) = a\begin{bmatrix} x+p & y+q \\ z+r & t+s \end{bmatrix} = \begin{bmatrix} a(x+p) & a(y+q) \\ a(z+r) & a(t+s) \end{bmatrix} \\
&= \begin{bmatrix} (ax+ap) & (ay+aq) \\ (az+ar) & (at+as) \end{bmatrix} = \begin{bmatrix} (ax) & (ay) \\ (az) & (at) \end{bmatrix} + \begin{bmatrix} (ap) & (aq) \\ (ar) & (as) \end{bmatrix} = aM_1 + aM_2
\end{aligned}
$$

ix. For each $M \in V$ and $a, b \in R$, $(a \cdot b) \cdot M = a \cdot (b \cdot M)$

$$
(a \cdot b) \cdot M = (a \cdot b) \cdot \begin{bmatrix} x & y \\ z & t \end{bmatrix} = \begin{bmatrix} abx & aby \\ abz & abt \end{bmatrix} = a \cdot \begin{bmatrix} bx & by \\ bz & bt \end{bmatrix} = a \cdot (b \cdot M)
$$

x. For each $M \in V$ and $1 \in R$,

$$
1 \cdot M = 1 \cdot \begin{bmatrix} x & y \\ z & t \end{bmatrix} = \begin{bmatrix} 1 \cdot x & 1 \cdot y \\ 1 \cdot z & 1 \cdot t \end{bmatrix} = \begin{bmatrix} x & y \\ z & t \end{bmatrix} = M
$$

Example 2: Consider a set $V = P^3 = $ set of all polynomials of degree of at most three in variable 'x' and coefficients from R (set of real numbers). Verify whether V forms a vector space over field $K = R$ where addition and scalar multiplication of polynomials is defined as

For $f(x) = f(x) = a_0 + a_1x + a_2x^2 + a_3x^3$ and $g(x) = b_0 + b_1x + b_2x^2 + b_3x^3$ in V
Define $f(x) + g(x) = (a_0 + b_0) + (a_1 + b_1)x + (a_2 + b_2)x^2 + (a_3 + b_3)x^3$
and $k(f(x)) = (ka_0) + (ka_1)x + (ka_{2_2})x^2 + (ka_3)x^3$, $k \in K$

Solution:

i. Consider $f(x) = a_0 + a_1x + a_2x^2 + a_3x^3$, $g(x) = b_0 + b_1x + b_2x^2 + b_3x^3$ in V then

$a_0, a_1, a_2, a_3, b_0, b_1, b_2, b_3 \in R$ then

$$
\begin{aligned}
f(x) + g(x) &= (a_0 + a_1x + a_2x^2 + a_3x^3) + (b_0 + b_1x + b_2x^2 + b_3x^3) \\
&= (a_0 + b_0) + (a_1 + b_1)x + (a_2 + b_2)x^2 + (a_3 + b_3)x^3
\end{aligned}
$$

belongs to V
since $(a_0 + b_0), (a_1 + b_1), (a_2 + b_2), (a_3 + b_3) \in R$

ii. Consider $f(x) = a_0 + a_1x + a_2x^2 + a_3x^3$, $g(x) = b_0 + b_1x + b_2x^2 + b_3x^3$ in V then

$a_0, a_1, a_2, a_3, b_0, b_1, b_2, b_3 \in R$ then

$$f(x) + g(x) = (a_0 + b_0) + (a_1 + b_1)x + (a_2 + b_2)x^2 + (a_3 + b_3)x^3$$
$$= (b_0 + a_0) + (b_1 + a_1)x + (b_2 + a_2)x^2 + (b_3 + a_3)x^3$$
$$= g(x) + f(x)$$

Thus, $f(x) + g(x) = g(x) + f(x)$ for every $f(x)$, $g(x)$ in V

iii. Consider $f(x) = a_0 + a_1x + a_2x^2 + a_3x^3$, $g(x) = b_0 + b_1x + b_2x^2 + b_3x^3$ and $h(x) = c_0 + c_1x + c_2x^2 + c_3x^3$ in V then $a_0, a_1, a_2, a_3, b_0, b_1, b_2, b_3, c_0, c_1, c_2, c_3 \in R$ then

$$[f(x) + g(x)] + h(x) = [(a_0 + b_0) + (a_1 + b_1)x + (a_2 + b_2)x^2 + (a_3 + b_3)x^3]$$
$$+ c_0 + c_1x + c_2x^2 + c_3x^3$$
$$= [(a_0 + b_0 + c_0) + (a_1 + b_1 + c_1)x + (a_2 + b_2 + c_2)x^2$$
$$+ (a_3 + b_3 + c_3)x^3]$$

$$f(x) + [g(x) + h(x)] = a_0 + a_1x + a_2x^2 + a_3x^3$$
$$+ [(b_0 + c_0) + (b_1 + c_1)x + (b_2 + c_2)x^2 + (b_3 + c_3)x^3]$$
$$= [(a_0 + b_0 + c_0) + (a_1 + b_1 + c_1)x + (a_2 + b_2 + c_2)x^2$$
$$+ (a_3 + b_3 + c_3)x^3]$$

Thus, $[f(x) + g(x)] + h(x) = f(x) + [g(x) + h(x)]$ for every $f(x)$, $g(x)$, $h(x)$ in V

iv. For every $f(x) = a_0 + a_1x + a_2x^2 + a_3x^3$ in V there is a zero vector $O(x) = 0 + 0x + 0x^2 + 0x^3$ in

V such that
$$f(x) + O(x) = (a_0 + a_1x + a_2x^2 + a_3x^3) + (0 + 0x + 0x^2 + 0x^3)$$
$$= (a_0 + 0) + (a_1 + 0)x + (a_2 + 0)x^2 + (a_3 + 0)x^3$$
$$= a_0 + a_1x + a_2x^2 + a_3x = f(x)$$

v. For every $f(x) = a_0 + a_1x + a_2x^2 + a_3x^3$ in V there is a inverse vector $-f(x) = (-a_0) + (-a_1)x + (-a_2)x^2 + (-a_3)x^3$ in V such that

$$f(x) + (-f(x)) = (a_0 + a_1x + a_2x^2 + a_3x^3) + ((-a_0) + (-a_1)x + (-a_2)x^2 + (-a_3)x^3)$$
$$= (a_0 + (-a_0)) + (a_1 + (-a_1))x + (a_2 + (-a_2))x^2 + (a_3 + (-a_3))x^3)$$
$$= = 0 + 0x + 0x^2 + 0x^3 = O(x)$$

vi. For every $f(x) = a_0 + a_1x + a_2x^2 + a_3x^3$ in V and $r \in R$, $aM_1 \in V$,
$r.f(x) = r.(a_0 + a_1x + a_2x^2 + a_3x^3) = (ra_0 + ra_1x + ra_2x^2 + ra_3x^3)$ belongs to V

since $ra_0, ra_1, ra_2, ra_3 \in R$

vii. For each $f(x) = a_0 + a_1x + a_2x^2 + a_3x^3 \in V$ and $r, s \in R$, then

$$
\begin{aligned}
(r + s)f(x) &= (r + s)(a_0 + a_1x + a_2x^2 + a_3x^3) \\
&= ((r + s)a_0 + (r + s)a_1x + (r + s)a_2x^2 + (r + s)a_3x^3) \\
&= (ra_0 + ra_1x + ra_2x^2 + ra_3x^3) + (sa_0 + sa_1x + sa_2x^2 + sa_3x^3) \\
&= rf(x) + sf(x)
\end{aligned}
$$

viii. Consider $f(x) = a_0 + a_1x + a_2x^2 + a_3x^3$, $g(x) = b_0 + b_1x + b_2x^2 + b_3x^3$ in V and $r \in R$,

$$
\begin{aligned}
r(f(x) + g(x)) &= r[(a_0 + b_0) + (a_1 + b_1)x + (a_2 + b_2)x^2 + (a_3 + b_3)x^3] \\
&= (ra_0 + rb_0) + (ra_1 + rb_1)x + (ra_2 + rb_2)x^2 + (ra_3 + rb_3)x^3 \\
&= (ra_0) + (ra_1)x + (ra_2)x^2 + (ra_3)x^3 + (rb_0) + (rb_1)x + (rb_2)x^2 \\
&\quad + (rb_3)x^3
\end{aligned}
$$

ix. For each $f(x) = a_0 + a_1x + a_2x^2 + a_3x^3$ and $r, s \in R$, $(r \cdot s) \cdot f(x) = r \cdot (s \cdot f(x))$

$$
\begin{aligned}
(r.s.)f(x) &= rs(a_0 + a_1x + a_2x^2 + a_3x^3) = (rsa_0 + rsa_1x + rsa_2x^2 + rsa_3x^3) \\
&= r(sa_0 + sa_1x + sa_2x^2 + sa_3x^3) = r(sf(x))
\end{aligned}
$$

x. For each $f(x) \in V$ and $1 \in R$,

$1.f(x) = 1.(a_0 + a_1x + a_2x^2 + a_3x^3) = a_0 + a_1x + a_2x^2 + a_3x^3 = f(x) \in V$

Exercise:

1. Consider a set $V = P^n =$ set of all polynomials of degree atmost 'n' in variable 'x' and coefficients from R (set of real numbers). Verify whether V forms a vector space over field R where addition and scalar multiplication of polynomials is defined as

$$f(x) + g(x) = (a_0 + b_0) + (a_1 + b_1)x + (a_2 + b_2)x^2 + (a_3 + b_3)x^3 + \ldots\ldots$$

$$+(a_n + b_n)x^n$$

$k(f(x)) = (ka_0) + (ka_1)x + (ka_2)x^2 + (ka_3)x^3 + \ldots\ldots +(ka_n)x^n, k \in R$, where
$f(x) = a_0 + a_1x + a_2x^2 + a_3x^3 + \ldots\ldots +a_nx^n$ and
$g(x) = b_0 + b_1x + b_2x^2 + b_3x^3 + \ldots\ldots +b_nx^n$ in V,

(Solution: $V = P^n =$ set of all polynomials of degree atmost 'n' in variable 'x' and coefficients from R is a vector space over R)

2. Consider a set $V = \left\{ M = \begin{bmatrix} a & 0 \\ b & c \end{bmatrix}, a, b, c > 0 \right\}$ set of lower triangular

matrices of order 2×2. Verify whether V forms a vector spaces over R (set of real numbers) under the usual operation of matrix addition and scalar multiplication.

(Solution: It is not a vector space. [Hint: additive inverse does not exist.])

2.1.1.1 Subspaces

Definition: Let V be a vector space over a field K and let W be a subset of V. Then W is a subspace of V if W is itself a vector space over K with respect to the operations of vector addition and scalar multiplication on V.

Also, alternately, we can define subspace W of vector space V as follows:

Suppose W is a subset of a vector space V. Then W is subspace of V if the following two conditions hold:

   i. $u + v \in W$, $u, v \in W$
  ii. There is a vector (called the zero vector) 0 in W such that
     $u + 0 = u$ $u \in W$
  iii. For each $u \in W$ and $a \in R$, $au \in W$ [1]

The above properties can be combined in a single statement as

For each $u, v \in W$ and $a, b \in R$, $au + bv \in W$

Example 1: Consider $V = R^3$ and W is subset of V consisting of all vectors in V such that each element in vector is equal. $W = \{(x, y, z): x = y = z\}$. Verify whether W is subspace of V.

Solution: Here $V = \{(x, y, z): x, y, z \in R\}$ and $W = \{(x, y, z): x = y = z\}$

  i. Clearly $0 = (0, 0, 0)$ belongs to W
  ii. Consider $u = (a, a, a)$, $v = (b, b, b) \in W$ then

$$u + v = (a, a, a) + (b, b, b) = (a + b, a + b, a + b) \in W$$

iii. Consider $u = (a, a, a) \in W$ and $r \in R$ then

$$ru = r(a, a, a) = (ra, ra, ra) \in W$$

Thus, W is subspace of vector space V over R.

Example 2: Consider vector space $V = M_{2 \times 2}$ set of matrices of order 2×2 whose entries belong to R (set of real numbers) over field K and W is a set of lower triangular matrices of order 2×2 whose entries belong to R (set of real numbers). Verify whether W is subspace of vector space V.

Solution: $V = \left\{ M = \begin{bmatrix} a & b \\ c & d \end{bmatrix}, a, b, c, d \in R \right\}$ of matrices of order 2×2 forms a vector spaces over a field R

$W = \left\{ M = \begin{bmatrix} a & 0 \\ b & c \end{bmatrix}, a, b, c \in R \right\}$ set of lower triangular matrices of order

2×2 is subset of V

i. Clearly $0_{2 \times 2} = \begin{bmatrix} 0 & 0 \\ 0 & 0 \end{bmatrix}$ belongs to W

ii. Consider $M_1 = \begin{bmatrix} a & 0 \\ c & d \end{bmatrix} \in W$, $M_2 = \begin{bmatrix} x & 0 \\ z & t \end{bmatrix} \in W$, then

$$M_1 + M_2 = \begin{bmatrix} a & 0 \\ c & d \end{bmatrix} + \begin{bmatrix} x & 0 \\ z & t \end{bmatrix} = \begin{bmatrix} a+x & 0 \\ c+z & d+t \end{bmatrix} \in W$$

iii. For each $M = \begin{bmatrix} a & 0 \\ c & d \end{bmatrix} \in W$, and $r \in R$,

$$r \cdot M = r \cdot \begin{bmatrix} a & 0 \\ c & d \end{bmatrix} = \begin{bmatrix} r \cdot a & 0 \\ r \cdot c & r \cdot d \end{bmatrix} = \begin{bmatrix} ra & 0 \\ rc & rd \end{bmatrix} \in W$$

Thus, W is subspace of vector space V over R.

Exercise:

1. Consider vector space $V = M_{2 \times 2}$ set of matrices of order 2×2 whose entries belong to R (set of real numbers) over field K and W is a set of upper triangular matrices of order 2×2 whose entries belong to R (set of real numbers). Verify whether W is subspace of vector space V.

(Solution: W is a subspace of a vector space V)

2. Consider $V = R^3$ and W is subset of V consisting all vectors in V such that the sum of each element in vector is equal to zero. $W = \{(x, y, z): x + y + z = 1\}$ Verify whether W is subspace of V.

(Solution: W is a not subspace of a vector space V.)

2.1.1.2 Affine Subspaces

Let V be a vector space, $v_0 \in V$, and $U \subseteq V$ a subspace. Then the subset $L = v_0 + U = \{v_0 + u: u \in U\} = \{v \in V:$ there exists $u \in U$ such that $v = v_0 + u\} \subseteq$ V is called affine subspace or linear manifold of V. U is called the direction or direction space, and v_0 is called the support point. Affine subspace is referred as a hyperplane if its dimension is one less than the dimension of vector space.

2.1.2 VECTOR NORM

A vector norm is the length of the vector also referred as the magnitude of the vector. This describes the extent of the vector in space.

- Let $x = (x_1, x_2 \ldots \ldots x_n)$ be a vector in R^n then p- norm of a vector is defined as

    $L^p = \|x\|_p = (\sum_{i=1}^{n} |x_i|^p)^{1/p}$ where p = 1, 2, 3, … .It is also referred as L^p norm of a vector.

    In particular, for p = 1, L^1 norm is defined as $L^1 = \|x\|_1 = (\sum_{i=1}^{n} |x_i|)$

    L^1 norm is also called as Manhattan distance.

    For p = 2, L^2 norm is defined as $L^2 = \|x\|_2 = (\sum_{i=1}^{n} |x_i|^2)^{1/2}$

    L^2 norm is also called the Euclidean norm.

    For p = ∞, L^∞ norm is defined as $L^\infty = \|x\|_\infty = \max_i |x_i|$

    L^∞ norm is also called the max norm or sup norm.

1. Find the L^1 norm of the vector $x = (1, -2, -3, 5)$

    $$L^1 = \|x\|_1 = \left(\sum_{i=1}^{n} |x_i|\right) = |1| + |-2| + |-3| + |5| = 1 + 2 + 3 + 5 = 11$$

2. Find the L^2 norm of the vector $x = (7, -1, -3, 2)$

    $$L^2 = \|x\|_2 = \left(\sum_{i=1}^{n} |x_i|^2\right)^{1/2} = \sqrt{(7)^2 + (-1)^2 + (-3)^2 + (2)^2} = \sqrt{63}$$

3. Find the L^∞ norm of the vector $x = (12, -11, -8, 6)$

    $$L^\infty = \|x\|_\infty = \max_i |x_i| = \max(|12|, |-11|, |-8|, |6|) = \max(12, 11, 8, 6) = 12$$

2.1.3 Linearly Dependency and Independency of Vectors

A linear combination of vectors $x_1, x_2, x_3. ....x_n$ in V with coefficients $a_1, a_2, a_3............ ..a_n$ in K is a vector of the form $a_1x_1 + a_2x_2 + a_3x_3 + ...+a_nx_n$.

A linear combination of vectors $a_1x_1 + a_2x_2 + a_3x_3 + ...+a_nx_n$ is called trivial if all coefficients $a_1, a_2, a_3............ ..a_n$ are zero.

A linear combination of vectors $a_1x_1 + a_2x_2 + a_3x_3 + ...+a_nx_n$ is called nontrivial if at least one of the coefficients $a_1, a_2, a_3............ ..a_n$ is nonzero.

Linearly independent vectors

The vectors $x_1, x_2, x_3. ....x_n$ are linearly independent if the linear combination of vectors $a_1x_1 + a_2x_2 + a_3x_3 + ...+a_nx_n$ is trivial.

Linearly dependent vectors

The vectors $x_1, x_2, x_3. ....x_n$ are linearly independent if the linear combination of vectors $a_1x_1 + a_2x_2 + a_3x_3 + ...+a_nx_n$ is not trivial.

Orthogonal Vectors Two vectors, x_1, x_2, are orthogonal if the dot product of any of these two vectors is zero.

Example 1: Examine whether the following vectors $(2, -1, 3, 2)$, $(1, 3, 4, 2)$ and $(3, -5, 2, 2)$ from $V = R^4$ are linear dependent or independent. If dependent, find the relation between them.

Solution: Let $x_1 = (2, -1, 3, 2)$ $x_2 = (1, 3, 4, 2)$ $x_3 = (3, -5, 2, 2)$
Consider $c1 \times 1 + c2 \times 2 + c3 \times 3 = 0... ... ..(1)$

$$c_1(2, -1, 3, 2) + c_2(1, 3, 4, 2) + c_3(3, -5, 2, 2) = 0$$
$$2c_1 + c_2 + 3c_3 = 0$$
$$- c_1 + 3c_2 - 5c_3 = 0$$
$$3c_1 + 4c_2 + 2c_3 = 0$$
$$2c_1 + 2c_2 + 2c_3 = 0$$

We can write it in matrix form as $AX = Z$,
$$\begin{bmatrix} 2 & 1 & 3 \\ -1 & 3 & -5 \\ 3 & 4 & 2 \\ 2 & 2 & 2 \end{bmatrix} \begin{bmatrix} c_1 \\ c_2 \\ c_3 \end{bmatrix} = \begin{bmatrix} 0 \\ 0 \\ 0 \\ 0 \end{bmatrix}$$

Consider $\left(A|0 \right) = \begin{bmatrix} 2 & 1 & 3 & | & 0 \\ -1 & 3 & -5 & | & 0 \\ 3 & 4 & 2 & | & 0 \\ 2 & 2 & 2 & | & 0 \end{bmatrix}$

Reducing the matrix to echelon form by performing row transformations we get

$$\frac{1}{2}R_4\sim \begin{bmatrix} 2 & 1 & 3 & | & 0 \\ -1 & 3 & -5 & | & 0 \\ 3 & 4 & 2 & | & 0 \\ 1 & 1 & 1 & | & 0 \end{bmatrix}, \quad R_{14}\sim \begin{bmatrix} 1 & 1 & 1 & | & 0 \\ -1 & 3 & -5 & | & 0 \\ 3 & 4 & 2 & | & 0 \\ 2 & 1 & 3 & | & 0 \end{bmatrix}$$

$$R_2 + R_1, R_3 - 3R_1, R_4 - 2R_1 \sim \begin{bmatrix} 1 & 1 & 1 & | & 0 \\ 0 & 4 & -4 & | & 0 \\ 0 & 1 & -1 & | & 0 \\ 0 & -1 & 1 & | & 0 \end{bmatrix}, \quad \frac{1}{4}R_2 \sim \begin{bmatrix} 1 & 1 & 1 & | & 0 \\ 0 & 1 & -1 & | & 0 \\ 0 & 1 & -1 & | & 0 \\ 0 & -1 & 1 & | & 0 \end{bmatrix}$$

$$R_3 - R_2, R_4 + R_2 \sim \begin{bmatrix} 1 & 1 & 1 & | & 0 \\ 0 & 1 & -1 & | & 0 \\ 0 & 0 & 0 & | & 0 \\ 0 & 0 & 0 & | & 0 \end{bmatrix}$$

A matrix is in echelon form rank (A) = rank (A,0) = 2 < *number of unknowns*

∴ system is consistent and has non − trivial solution

Thus, all c_i 's are not equal to zero. Hence, vectors are linearly dependent. To find a relation between the vectors, consider the equation

$$c_1 + c_2 + c_3 = 0$$
$$c_2 - c_3 = 0$$

There are two equations in three unknowns. Therefore, we can choose the 3 − 2 = 1 parameter
Let us choose the parameter k and set $c_3 = k$ implies that $c_2 = k$ and $c_1 = -k - k$

$$c_1 = -2k$$

Substitute c_1, c_2, c_3 in equation (1).

$$-2kx_1 + kx_2 + kx_3 = 0,$$
$$-2x_1 + x_2 + x_3 = 0$$
$$2x_1 = x_2 + x_3$$

Example 2: Examine the linear dependence or independence for the given vectors and if dependent, find the relation between them: $x_1 = (1, -1, 2, 2);$ $x_2 = (2, -3, 2, -1);$ $x_3 = (-1, 2, -2, 3)$

Solution: Given vectors are $x_1 = (1, -1, 2, 2);$ $x_2 = (2, -3, 2, -1);$ $x_3 = (-1, 2, -2, 3)$
Consider $c_1x_1 + c_2x_2 + c_3x_3 = 0$... ... ..(1)

$$c_1(1, -1, 2, 2) + c_2(2, -3, 2, -1) + c_3(-1, 2, -2, 3) = 0$$
$$c_1 + 2c_2 - c_3 = 0$$
$$-c_1 - 3c_2 + 2c_3 = 0$$
$$2c_1 + 2c_2 - 2c_3 = 0$$
$$2c_1 - c_2 + 3c_3 = 0$$

We can write it in matrix form as AX = Z.

$$\begin{bmatrix} 1 & 2 & -1 \\ -1 & -3 & 2 \\ 2 & 2 & -2 \\ 2 & -1 & 3 \end{bmatrix} \begin{bmatrix} c_1 \\ c_2 \\ c_3 \end{bmatrix} = \begin{bmatrix} 0 \\ 0 \\ 0 \\ 0 \end{bmatrix}$$

Consider $A = \begin{bmatrix} 1 & 2 & -1 \\ -1 & -3 & 2 \\ 2 & 2 & -2 \\ 2 & -1 & 3 \end{bmatrix}$. Reducing the

$$R_2 + R_1; \; R_3 - 2R_1; \; R_4 - 2R_1 \sim \begin{bmatrix} 1 & 2 & -1 \\ 0 & -1 & 1 \\ 0 & -2 & 0 \\ 0 & -5 & 5 \end{bmatrix}$$

$$R_4 - 5R_2 \sim \begin{bmatrix} 1 & 2 & -1 \\ 0 & -1 & 1 \\ 0 & -2 & 0 \\ 0 & 0 & 0 \end{bmatrix}$$

$$R_3 - 2R_2 \sim \begin{bmatrix} 1 & 2 & -1 \\ 0 & -1 & 1 \\ 0 & 0 & -2 \\ 0 & 0 & 0 \end{bmatrix}$$

The matrix is in echelon form.

rank(A) = rank(A , 0) = 3 = number of unknowns
∴ system is consistent and has trivial solution
∴ vectors are linearly independent

Example 3: Examine whether the following vectors $x = (3, 0, 4)$, $y = (-4, 0, 3)$, $z = (0, 9, 0)$ are orthogonal.

Solution:

$$x \bullet y = (3) \cdot (-4) + (0) \cdot (0) + (4) \cdot (3) = 0$$

$$x \bullet z = (3) \cdot (0) + (0) \cdot (9) + (4) \cdot (0) = 0$$

$$y \bullet z = (-4) \cdot (0) + (0) \cdot (9) + (3) \cdot (0) = 0$$

Since the dot product of these vectors is zero, they are mutually orthogonal.

Example 4: Examine whether the following vectors $x = (1, 0, 0)$, $y = (0, 0, 1)$, $z = (0, 1, 0)$ are orthogonal.

Solution:

$$x \bullet y = (1) \cdot (0) + (0) \cdot (0) + (0) \cdot (1) = 0$$

$$x \bullet z = (1) \cdot (0) + (0) \cdot (1) + (0) \cdot (0) = 0$$

$$y \bullet z = (0) \cdot (0) + (0) \cdot (1) + (1 \cdot (0) = 0$$

Since the dot product of these vectors is zero, they are mutually orthogonal.
Magnitude of vector

Exercise:

1. Examine whether the following vectors are linear dependent. Find the relation between them; if dependent, $x = (1, 2, 4)$, $y = (2, -1, 3)$, $r = (0, 8, 5)$, $s = (-3, 7, -2)$.

   Solution: Vectors are linearly dependent. Relation: $x + y + s = r$

2. Examine whether the following vectors are linear dependent. Find the relation between them; if dependent, $x = (1, 2, 0)$, $y = (0, -1, 3)$, $z = (3, 8, 0)$

   Solution: Vectors are linearly independent.

3. Examine whether the following vectors $x = (1, 0, 0)$, $y = (0, 0, 1)$, $z = (0, 1, 0)$ are orthogonal.

   Solution: Vectors are orthogonal.

4. Examine whether the following vectors $x = (2, 0, 4)$, $y = (5, 3, 1)$, $z = (0, 7, -2)$ are orthogonal.

   Solution: Vectors are not orthogonal.

2.1.4 BASIS AND DIMENSIONS OF VECTORS SPACES

Spanning set of a vector space V

A set of vectors $x_1, x_2, x_3. ....x_n$ in V is said to be a spanning set if every vector in V can be expressed as the linear combination of $x_1, x_2, x_3. ....x_n$.

Example: Consider a vector space $V = R^3$.

Spanning set of $V = e_1 = (1, 0, 0),$ $e_2 = (0, 1, 0),$ $e_3 = (0, 0, 1)$

Any vector in R^3 (x,y,z) can be expressed as a linear combination of e_1, e_2, e_3.

$$(x, y, z) = x(1, 0, 0) + y(0, 1, 0) + z(0, 0, 1)$$

Basis of a vector space V

A set of vectors $S = \{x_1, x_2, x_3..............x_n\}$ in V is a basis of V if it has the following two properties:

  i. $S = \{x_1, x_2, x_3..............x_n\}$ is linearly independent.
  ii. $S = \{x_1, x_2, x_3..............x_n\}$ spans the vector space V.

Dimension of a vector space V

Dimension of a vector space V is the number of elements in the basis of vector space V.

Example 1: Consider a vector space $V = \left\{ M = \begin{bmatrix} a & b \\ c & d \end{bmatrix}, a, b, c, d \in R \right\}$ of matrices of order 2 × 2. Find the basis and dimension of vector space V.

Since every element of vector space V can be expressed as a linear combination of

$$E_1 = \begin{bmatrix} 1 & 0 \\ 0 & 0 \end{bmatrix}, \quad E_2 = \begin{bmatrix} 0 & 1 \\ 0 & 0 \end{bmatrix}, \quad E_3 = \begin{bmatrix} 0 & 0 \\ 1 & 0 \end{bmatrix}, \quad E_4 = \begin{bmatrix} 0 & 0 \\ 0 & 1 \end{bmatrix}$$

$$M = \begin{bmatrix} a & b \\ c & d \end{bmatrix} = a \begin{bmatrix} 1 & 0 \\ 0 & 0 \end{bmatrix} + b \begin{bmatrix} 0 & 1 \\ 0 & 0 \end{bmatrix} + c \begin{bmatrix} 0 & 0 \\ 1 & 0 \end{bmatrix} + d \begin{bmatrix} 0 & 0 \\ 0 & 1 \end{bmatrix}$$

Basis of vector space $V = \left\{ E_1 = \begin{bmatrix} 1 & 0 \\ 0 & 0 \end{bmatrix}, E_2 = \begin{bmatrix} 0 & 1 \\ 0 & 0 \end{bmatrix}, E_3 = \begin{bmatrix} 0 & 0 \\ 1 & 0 \end{bmatrix}, E_4 = \begin{bmatrix} 0 & 0 \\ 0 & 1 \end{bmatrix} \right\}$

Dimension of vector space V = no of element in the basis of vector space V = 4.

Exercise:

  1. Consider a vector space $V = \left\{ M = \begin{bmatrix} a & 0 \\ b & c \end{bmatrix}, a, b, c \in R \right\}$ of matrices of order 2 × 2.

Find the basis and dimension of vector space V.

Solution: Basis of vector space $V = \left\{ E_1 = \begin{bmatrix} 1 & 0 \\ 0 & 0 \end{bmatrix}, E_2 = \begin{bmatrix} 0 & 0 \\ 0 & 1 \end{bmatrix}, E_3 = \begin{bmatrix} 0 & 0 \\ 1 & 0 \end{bmatrix} \right\}$,

dimension of vector space V = no of element in the basis of vector space V = 3.

2. Consider P^3 = set of all polynomials of degree atmost '3' in variable 'x' and coefficients from R.

Does the set $B = (1, x + 2, x^2 + 3)$ form a basis of vector space V?

Solution: Yes

2.1.5 LINEAR TRANSFORMATION (LINEAR MAPPING)

Consider two vector spaces V and W over field K. A mapping $T: V \rightarrow W$ is called linear if

$T(au + bv) = aT(u) + bT(v)$ for all $u, v \in V$ and for all $a, b \in K$.

Image of $T = \{w \in W$ for which there exist $v \in V$ such that $T(u) = w\}$

Kernel of $T = \{v \in V$ such that $T(v) = 0$ where $0 \in W\}$

Rank-Nullity Theorem

For vector spaces V, W, and if $T: V \rightarrow W$ is a linear transformation, then

$$\text{rank}(T) + \text{Nullity}(T) = dim(V)$$

where rank(T) = dim (Image(T)) and Nullity(T) = dim(Ker(T))

$$\dim(\text{Im}T) + \dim(\text{Ker}T) = \dim(V).$$

Example 1: Consider a vector space P_3 over a field K and a derivative map $T: P_3 \rightarrow P_3$ defined as $T(f(x)) = \frac{d}{dx}f(x)$ where P_3 is a set of polynomials of order almost 3 Verify Rank-Nullity Theorem.

Solution: Derivative map $T: P_3 \rightarrow P_3$ is linear since $T(af(x) + bg(x)) = \frac{d}{dx}(af(x) + bg(x)) = a\left(\frac{d}{dx}f(x)\right) + b\left(\frac{d}{dx}g(x)\right)$ for all $f(x), g(x) \in P_3$ and for all $a, b \in K$

Consider the basis of P_3: $\{1,x,x^2,x^3\}$ since every element of vector space P_3 can be expressed as linear combination of $1,x,x^2,x^3$.

$$T(1) \quad = \quad \frac{d}{dx}(1) = 0 = 0(1) + 0(x) + 0(x^2) + 0(x^3)$$

$$T(x) \quad = \quad \frac{d}{dx}(x) = 1 = 1(1) + 0(x) + 0(x^2) + 0(x^3)$$

$$T(x^2) \quad = \quad \frac{d}{dx}(x^2) = 2x = 0(1) + 2(x) + 0(x^2) + 0(x^3)$$

$$T(x^3) \quad = \quad \frac{d}{dx}(x^3) = 3x^2 = 0(1) + 0(x) + 3(x^2) + 0(x^3)$$

Matrix of transformation $A = \begin{bmatrix} 0 & 1 & 0 & 0 \\ 0 & 0 & 2 & 0 \\ 0 & 0 & 0 & 3 \\ 0 & 0 & 0 & 0 \end{bmatrix}$

$$T(v) = A(v)$$

$A(v)$ is matrix representation of lineartransformation $T(v)$.

The matrix is in echelon form:

dim (Image(T)) = rank(A) = number of non zero rows in echelon form = 3

dim(ker(T)) = (Nullity(A) = $\eta(A)$ = number of zero rows in echelon form = 1

dim(KerT) + dim(ImT) = dim(V).

rank(A) + Nullity (A) = 4

Example 2: If A is a 4 × 6 matrix with rank 2, what is the nullity of A?

Solution: Since the nullity is the difference between the number of columns of A and the rank of A, the nullity of this matrix is 6 − 4 = 2.

Exercise. If A is a 5 × 3 matrix with a nullity 2, what is the rank of A? [**Solution:** Rank of A = 1.]

2.1.6   EIGENVALUE AND EIGENVECTOR OF A LINEAR TRANSFORMATION

Consider a linear transformation T from a vector space V over a field F into itself $T: V \rightarrow V$ and v is a nonzero vector in V, then v is an eigenvector of T if $T(v)$ is a scalar multiple of v.

This can be written as $T(v) = \lambda v$,

where λ is a scalar in field F known as the eigenvalue, characteristic value, or characteristic root associated with eigenvector v.

If the vector space V is finite dimensional, then linear transformation can be expressed in the form of an n by n matrix A, then the eigenvalue equation for a linear transformation can be rewritten as $A(v) = \lambda v$.

The set of all eigenvectors of a linear transformation, each paired with its corresponding eigenvalue, is called the **eigensystem** of that transformation.

Example 1: Find eigenvalues and eigenvectors of following matrix A = $\begin{bmatrix} 1 & 1 & 1 \\ 0 & 2 & 1 \\ 0 & 0 & 3 \end{bmatrix}$.

Solution: We have $A = \begin{bmatrix} 1 & 1 & 1 \\ 0 & 2 & 1 \\ 0 & 0 & 3 \end{bmatrix}$

as the matrix is an upper triangular matrix and the eigenvalues are main diagonal entries.

$$\therefore \text{ Eigenvalue of matrix } A, \; \lambda_1 = 1, \; \lambda_2 = 2, \; \lambda_3 = 3$$

To find eigenvectors, consider $(A - \lambda I)X = 0$,

where the eigenvector is $X = \begin{bmatrix} x \\ y \\ z \end{bmatrix}$ $\begin{bmatrix} 1-\lambda & 1 & 1 \\ 0 & 2-\lambda & 1 \\ 0 & 0 & 3-\lambda \end{bmatrix}\begin{bmatrix} x \\ y \\ z \end{bmatrix} = \begin{bmatrix} 0 \\ 0 \\ 0 \end{bmatrix}$ .....(1)

Put $\lambda = 1$ $(A - I)X = 0$ $\begin{bmatrix} 0 & 1 & 1 \\ 0 & 1 & 1 \\ 0 & 0 & 2 \end{bmatrix}\begin{bmatrix} x \\ y \\ z \end{bmatrix} = \begin{bmatrix} 0 \\ 0 \\ 0 \end{bmatrix}$

$$2z = 0$$
$$y + z = 0$$

let $x = t$ be the parameter

$$\begin{bmatrix} x \\ y \\ z \end{bmatrix} = \begin{bmatrix} t \\ 0 \\ 0 \end{bmatrix} = t\begin{bmatrix} 1 \\ 0 \\ 0 \end{bmatrix}$$

$$\therefore \text{ eigen vectors corresponding to } \lambda = 1 = \begin{bmatrix} 1 \\ 0 \\ 0 \end{bmatrix}$$

Put $\lambda = 2$ in (1)

$$\begin{bmatrix} -1 & 1 & 1 \\ 0 & 0 & 1 \\ 0 & 0 & 1 \end{bmatrix}\begin{bmatrix} x \\ y \\ z \end{bmatrix} = \begin{bmatrix} 0 \\ 0 \\ 0 \end{bmatrix}$$

$$z = 0$$
$$\therefore -x + y + z = 0$$
$$\therefore -x + y = 0$$

This is one equation with two unknowns:

$$\therefore \text{ we can choose } 2 - 1 = 1 \text{ parameter}$$

$$\text{let } y = t \text{ be the parameter}$$

$$-x + t = 0$$

$$\Rightarrow x = t$$

$$\begin{bmatrix} x \\ y \\ z \end{bmatrix} = \begin{bmatrix} t \\ t \\ 0 \end{bmatrix} = t \begin{bmatrix} 1 \\ 1 \\ 0 \end{bmatrix}$$

$$\therefore \text{ eigen vectors corresponding to } \lambda = 2 = \begin{bmatrix} 1 \\ 1 \\ 0 \end{bmatrix}$$

Put $\lambda = 3$ in (1)

$$\begin{bmatrix} -2 & 1 & 1 \\ 0 & -1 & 1 \\ 0 & 0 & 0 \end{bmatrix} \begin{bmatrix} x \\ y \\ z \end{bmatrix} = \begin{bmatrix} 0 \\ 0 \\ 0 \end{bmatrix}$$

$$-2x + y + z = 0$$

$$-y + z = 0$$

Two equations with three unknowns. We can choose the 3−2=1 parameter. Let $z = t$ be the parameter

$$-y + t = 0 \therefore y = t - 2x + t + t = 0 \; 2x = 2t \therefore x = t$$

$$\begin{bmatrix} x \\ y \\ z \end{bmatrix} = \begin{bmatrix} t \\ t \\ t \end{bmatrix} = t \begin{bmatrix} 1 \\ 1 \\ 1 \end{bmatrix}$$

Eigenvector for $\lambda = 3$ is $\begin{bmatrix} 1 \\ 1 \\ 1 \end{bmatrix}$.

Exercise: Find the eigenvalues and eigenvectors for $A = \begin{bmatrix} 0 & 2 & 0 \\ 3 & -2 & 3 \\ 0 & 3 & 0 \end{bmatrix}$.

Solution: Eigenvalues are 0, 3, −5.

Eigen vectors are $\lambda = 0$ is $\begin{bmatrix} 1 \\ 0 \\ -1 \end{bmatrix}$, $\lambda = 3$ is $\begin{bmatrix} 2 \\ 3 \\ 3 \end{bmatrix}$, $\lambda = -5$ is $\begin{bmatrix} 2 \\ -5 \\ 3 \end{bmatrix}$.

2.1.7 MATRIX FACTORIZATION

Matrix decompositions are methods that reduce a matrix into constituent parts that make it easier to calculate more complex matrix operations. Matrix decomposition methods are also called matrix factorization methods. Three simple and widely used matrix decomposition methods are the LU matrix decomposition, Cholesky's decomposition, and the QR matrix decomposition.

2.1.7.1 LU Decomposition

LU decomposition of a matrix is the factorization of a given square matrix into two triangular matrices, one upper triangular matrix and one lower triangular matrix, such that the product of these two matrices gives the original matrix.

Square matrix A can be decomposed into two square matrices, L and U, such that $A = L\,U$ where U is an upper triangular matrix and L is a lower triangular matrix with diagonal elements being equal to 1.

For $A = \begin{bmatrix} a_{11} & a_{12} & a_{13} \\ a_{21} & a_{22} & a_{23} \\ a_{31} & a_{32} & a_{33} \end{bmatrix}$, we have $L = \begin{bmatrix} 1 & 0 & 0 \\ l_{21} & 1 & 0 \\ l_{31} & l_{32} & 1 \end{bmatrix}$

and $U = \begin{bmatrix} u_{11} & u_{12} & u_{13} \\ 0 & u_{22} & u_{23} \\ 0 & 0 & u_{33} \end{bmatrix}$; such that $A = LU$

Example 1: Solve the system of equations by triangular factorization method:

$$2x_1 + 2x_2 + 3x_3 = 4$$
$$4x_1 - 2x_2 + x_3 = 9$$
$$x_1 + 5x_2 + 4x_3 = 3$$

Solution: Given system can be written in the matrix form as $AX = B$

where $A = \begin{bmatrix} 2 & 2 & 3 \\ 4 & -2 & 1 \\ 1 & 5 & 4 \end{bmatrix}$, $X = \begin{bmatrix} x_1 \\ x_2 \\ x_3 \end{bmatrix}$, $B = \begin{bmatrix} 4 \\ 9 \\ 3 \end{bmatrix}$

Now, we express $A = LU$

$$\begin{bmatrix} 2 & 2 & 3 \\ 4 & -2 & 1 \\ 1 & 5 & 4 \end{bmatrix} = \begin{bmatrix} 1 & 0 & 0 \\ l_{21} & 1 & 0 \\ l_{31} & l_{32} & 1 \end{bmatrix} \begin{bmatrix} u_{11} & u_{12} & u_{13} \\ 0 & u_{22} & u_{23} \\ 0 & 0 & u_{33} \end{bmatrix}$$

Multiplying the right-hand side of the equation of the elements in corresponding position, we get

i. $u_{11} = 2,\ u_{12} = 2,\ u_{13} = 3$

ii. $l_{21}u_{11} = 4 \Rightarrow l_{21} = \frac{4}{2} = 2,\ l_{31}u_{11} = 1 \Rightarrow l_{31} = \frac{1}{2}$

iii.
$$l_{21}u_{12} + u_{22} = -2 \Rightarrow u_{22} = -2 - (2)(2) = -6$$
$$l_{21}u_{13} + u_{23} = 1 \Rightarrow u_{23} = 1 - (2)(3) = -5$$

iv.
$$l_{31}u_{12} + l_{32}u_{22} = 5 \Rightarrow l_{32} = \frac{1}{6}\left[5 - \frac{1}{2}(2)\right] = -\frac{2}{3},$$
$$l_{31}u_{13} + l_{32}u_{23} + u_{33} = 4, u_{33} = 4 - \frac{1}{2}(3) - \left(-\frac{2}{3}\right)(-5) = -\frac{5}{6}$$

Thus, we have,

$$L = \begin{bmatrix} 1 & 0 & 0 \\ 2 & 1 & 0 \\ 1/2 & -2/3 & 1 \end{bmatrix}, \quad U = \begin{bmatrix} 2 & 2 & 3 \\ 0 & -6 & -5 \\ 0 & 0 & -5/6 \end{bmatrix} \dots\dots\dots(3)$$

Given system (1) can be written as

$$LUX = B$$

Let $UX = Z$ $\therefore LZ = B$, where $Z = [z_1, z_2, z_3]^T$
$LZ = B$ gives

$$\begin{bmatrix} 1 & 0 & 0 \\ 2 & 1 & 0 \\ 1/2 & -2/3 & 1 \end{bmatrix}\begin{bmatrix} z_1 \\ z_2 \\ z_3 \end{bmatrix} = \begin{bmatrix} 4 \\ 9 \\ 3 \end{bmatrix}$$

Which is equivalent to

$$z_1 = 4, \quad 2z_1 + z_2 = 9, \quad \frac{1}{2}z_1 - \frac{2}{3}z_2 + z_3 = 3$$

Using forward substitution, we obtain

$$z_1 = 4, \quad z_2 = 1, \quad z_3 = 5/3$$

Hence, the solution of original system is given by, $UX = Z$ as

$$\begin{bmatrix} 2 & 2 & 3 \\ 0 & -6 & -5 \\ 0 & 0 & -5/6 \end{bmatrix}\begin{bmatrix} x_1 \\ x_2 \\ x_3 \end{bmatrix} = \begin{bmatrix} 4 \\ 1 \\ 5/3 \end{bmatrix}$$

Which is equivalent to

$$2x_1 + 2x_2 + 3x_3 = 4, \quad -6x_2 - 5x_3 = 1, \quad -5/6x_3 = 5/3$$

Solving using backward substitution, we have

$$x_1 = -7/2, \quad x_2 = 3/2, \quad x_3 = -2$$

Exercise: Solve the following system of equations by LU decomposition:

$$2x + 3y + z = 9$$
$$x + 2y + 3z = 6$$
$$3x + y + 2z = 8$$

Solution: $x = \frac{35}{18}$, $y = \frac{29}{18}$, $z = \frac{5}{18}$.

2.1.7.2 Cholesky's Decomposition

The **Cholesky decomposition** or **Cholesky factorization** is a decomposition of a Hermitian, positive-definite matrix into the product of a lower triangular matrix and its conjugate transpose. The Cholesky decomposition is twice as efficient as the LU decomposition. The Cholesky decomposition of a Hermitian positive-definite matrix A is a decomposition of the form $A = LL^t$, where L is a lower triangular matrix with real and positive diagonal entries, and L^t denotes the conjugate transpose of L. Every Hermitian positive-definite matrix (and thus also every real-valued symmetric positive-definite matrix) has a unique Cholesky decomposition.

For $A = \begin{bmatrix} a_{11} & a_{12} & a_{13} \\ a_{21} & a_{22} & a_{23} \\ a_{31} & a_{32} & a_{33} \end{bmatrix}$, we have $L = \begin{bmatrix} l_{11} & 0 & 0 \\ l_{21} & l_{22} & 0 \\ l_{31} & l_{32} & l_{33} \end{bmatrix}$

$L^t = \begin{bmatrix} l_{11} & l_{21} & l_{31} \\ 0 & l_{22} & l_{32} \\ 0 & 0 & l_{33} \end{bmatrix}$ and; such that $A = LL^t$

Example 1: Solve the following system of equations by Cholesky's method:

$$2x_1 - x_2 = 1$$
$$- x_1 + 3x_2 + x_3 = 0$$
$$x_2 + 2x_3 = 0$$

Solution: Given system can be expressed in matrix form as AX = B,

Where $A = \begin{bmatrix} 2 & -1 & 0 \\ -1 & 3 & 1 \\ 0 & 1 & 2 \end{bmatrix}$, $X = \begin{bmatrix} x_1 \\ x_2 \\ x_3 \end{bmatrix}$, $B = \begin{bmatrix} 1 \\ 0 \\ 0 \end{bmatrix}$

Now we express $A = LL^T$

i.e $\begin{bmatrix} 2 & -1 & 0 \\ -1 & 3 & 1 \\ 0 & 1 & 2 \end{bmatrix} = \begin{bmatrix} l_{11} & 0 & 0 \\ l_{21} & l_{22} & 0 \\ l_{31} & l_{32} & l_{33} \end{bmatrix} \begin{bmatrix} l_{11} & l_{12} & l_{13} \\ 0 & l_{22} & l_{23} \\ 0 & 0 & l_{33} \end{bmatrix}$

Multiplying the right-hand side and equating the elements in corresponding position, we get

$$l_{11}^2 = 2 \quad \therefore \quad l_{11} = \sqrt{2}$$

$$l_{11}l_{21} = -1 \quad \therefore \quad l_{21} = -\frac{1}{\sqrt{2}}$$

$$l_{11}l_{31} = 0 \quad \therefore \quad l_{31} = 0$$

$$l_{21}^2 + l_{22}^2 = 3 \quad \therefore \quad l_{22} = \sqrt{\frac{5}{2}}$$

$$l_{21}l_{31} + l_{22}l_{32} = 1 \quad \therefore \quad l_{32} = \sqrt{\frac{2}{5}}$$

$$l_{31}^2 + l_{32}^2 + l_{33}^2 = 2 \quad \therefore \quad l_{32} = \sqrt{\frac{8}{5}}$$

Thus, we have

$$L = \begin{bmatrix} \sqrt{2} & 0 & 0 \\ -\frac{1}{\sqrt{2}} & \sqrt{\frac{5}{2}} & 0 \\ 0 & \sqrt{2/5} & \sqrt{8/5} \end{bmatrix}$$

The given system can be written as $LL^TX = B$.
Let $L^TX = Z \therefore LZ = B$ where $Z = [z_1 \ z_2 \ z_3]^T$.
Now consider $LZ = B$

$$\begin{bmatrix} \sqrt{2} & 0 & 0 \\ -1/\sqrt{2} & \sqrt{5/2} & 0 \\ 0 & \sqrt{2/5} & \sqrt{8/5} \end{bmatrix} \begin{bmatrix} z_1 \\ z_2 \\ z_3 \end{bmatrix} = \begin{bmatrix} 1 \\ 0 \\ 0 \end{bmatrix}$$

Which is equivalent to $\sqrt{2}z_1 = 1$, $\frac{-1}{\sqrt{2}}z_1 + \sqrt{\frac{5}{2}}z_2 = 0$, $\sqrt{\frac{2}{5}}z_2 + \sqrt{\frac{8}{5}}z_3 = 0$

Using forward substitution, we obtain $z_1 = \frac{1}{\sqrt{2}}$, $z_2 = \frac{1}{\sqrt{10}}$, $z_3 = \frac{-1}{\sqrt{40}}$.

Hence, the solution of the original system is given by $L^TX = Z$ as

$$\begin{bmatrix} \sqrt{2} & \frac{-1}{\sqrt{2}} & 0 \\ 0 & \sqrt{5/2} & \sqrt{2/5} \\ 0 & 0 & \sqrt{8/5} \end{bmatrix} \begin{bmatrix} x_1 \\ x_2 \\ x_3 \end{bmatrix} = \begin{bmatrix} \frac{1}{\sqrt{2}} \\ \frac{1}{\sqrt{10}} \\ \frac{-1}{\sqrt{40}} \end{bmatrix}$$

Which is equivalent to

$$\sqrt{2}\,x_1 - \frac{1}{\sqrt{2}}x_2 = \frac{1}{\sqrt{2}};\quad \sqrt{\frac{5}{2}}\,x_2 + \sqrt{\frac{2}{5}}\,x_3 = \frac{1}{\sqrt{10}};\quad \sqrt{\frac{8}{5}}\,x_3 = \frac{-1}{\sqrt{40}}$$

By backwards substitution, we get

$$x_1 = 5/8,\ x_2 = 1/4,\ x_3 = -1/8$$

Exercise: Solve the following system of equations by Cholesky's method:

$$
\begin{aligned}
4x_1 - 2x_2 \qquad\quad &= 0\\
-2x_1 + 4x_2 - x_3 &= 1\\
- x_2 + 4x_3 &= 0
\end{aligned}
$$

Solution: $x_1 = \frac{2}{11},\ x_2 = \frac{4}{11},\ x_3 = \frac{1}{11}$

2.1.7.3 QR Decomposition

The QR decomposition (also called the QR factorization) of a matrix is a decomposition of the matrix into an orthogonal matrix and a triangular matrix. A QR decomposition of a real square matrix A is a decomposition of A as $A = QR$, where Q is an orthogonal matrix (i.e., $Q'Q = QQ' = I$) and R is an upper triangular matrix. If A is nonsingular, then this factorization is unique. There are several methods for actually computing the QR decomposition. One of such method is the Gram-Schmidt process.

Gram-Schmidt process

Consider the Gram Schmidt procedure, with the vectors to be considered in the process as columns of the matrix A.

$$A = [a_1 \ \ a_2 \ \ a_3 \ \ a_4 \ \ a_5]$$

Then $u_1 = a_1,\ e_1 = \frac{u_1}{\|u_1\|}$

$$
\begin{aligned}
u_2 &= a_2 - \langle a_2.\ e_1\rangle e_1, & e_2 &= \frac{u_2}{\|u_2\|}\\
u_3 &= a_3 - \langle a_3.\ e_1\rangle e_1 - \langle a_3.\ e_2\rangle e_2, & e_3 &= \frac{u_3}{\|u_3\|}\\
u_4 &= a_4 - \langle a_4.\ e_1\rangle e_1 - \langle a_4.\ e_2\rangle e_2 - \langle a_4.\ e_3\rangle e_3, & e_4 &= \frac{u_4}{\|u_4\|}\\
u_5 &= a_5 - \langle a_5.\ e_1\rangle e_1 - \langle a_5.\ e_2\rangle e_2 - \langle a_5.\ e_3\rangle e_3 - \langle a_5.\ e_4\rangle e_4, & e_5 &= \frac{u_5}{\|u_5\|}
\end{aligned}
$$

Here $\|.\|$ is L_2 norm

The resulting QR factorization is

•

$$A = [a_1 \ a_2 \ a_3 \ a_4 \ a_5] = [e_1 \ e_2 \ e_3 \ e_4 \ e_5]. \begin{bmatrix} a_1e_1 & a_2e_1 & a_3e_1 & a_4e_1 & a_5e_1 \\ 0 & a_2e_2 & a_3e_2 & a_4e_2 & a_5e_2 \\ 0 & 0 & a_3e_3 & a_4e_3 & a_5e_3 \\ 0 & 0 & 0 & a_4e_4 & a_5e_4 \\ 0 & 0 & 0 & 0 & a_5e_5 \end{bmatrix} = QR$$

Example 1: Factorize a matrix $A = \begin{bmatrix} 1 & -1 & 0 \\ 2 & 0 & 0 \\ 2 & 2 & 1 \end{bmatrix}$ by the Gram-Schmidt process.

Solution: $a_1 = [1 \ 2 \ 2]$, $a_2 = [-1 \ 0 \ 2]$, $a_3 = [0 \ 0 \ 1]$,

$$u_1 = a_1 = [1 \ 2 \ 2], \quad e_1 = \frac{u_1}{\|u_1\|} = \frac{|1 \ 2 \ 2|}{3} = \begin{bmatrix} \frac{1}{3} & \frac{2}{3} & \frac{2}{3} \end{bmatrix}$$

$$u_2 = a_2 - \langle a_2. \ e_1\rangle e_1 = [-1 \ 0 \ 2] - \begin{bmatrix} \frac{1}{3} & \frac{2}{3} & \frac{2}{3} \end{bmatrix} = \begin{bmatrix} \frac{-4}{3} & \frac{-2}{3} & \frac{4}{3} \end{bmatrix},$$

$$e_2 = \frac{u_2}{\|u_2\|} = \begin{bmatrix} \frac{-2}{3} & \frac{-1}{3} & \frac{2}{3} \end{bmatrix}$$

$$u_3 = a_3 - \langle a_3. \ e_1\rangle e_1 - \langle a_3. \ e_2\rangle e_2$$

$$= [0 \ 0 \ 1] - \frac{2}{3}\begin{bmatrix} \frac{1}{3} & \frac{2}{3} & \frac{2}{3} \end{bmatrix} - \frac{2}{3}\begin{bmatrix} \frac{-2}{3} & \frac{-1}{3} & \frac{2}{3} \end{bmatrix} = \begin{bmatrix} \frac{2}{9} & \frac{-2}{9} & \frac{1}{9} \end{bmatrix},$$

$$e_3 = \frac{u_3}{\|u_3\|} = \begin{bmatrix} \frac{2}{3} & \frac{-2}{3} & \frac{1}{3} \end{bmatrix}$$

$$A = [e_1 \ e_2 \ e_3] \begin{bmatrix} a_1e_1 & a_2e_1 & a_3e_1 \\ 0 & a_2e_2 & a_3e_2 \\ 0 & 0 & a_3e_3 \end{bmatrix}. = QR$$

$$Q = [e_1 \ e_2 \ e_3] = \begin{bmatrix} \frac{1}{3} & -\frac{2}{3} & \frac{2}{3} \\ \frac{2}{3} & -\frac{1}{3} & \frac{-2}{3} \\ \frac{2}{3} & \frac{2}{3} & \frac{1}{3} \end{bmatrix}.$$

$$R = \begin{bmatrix} a_1e_1 & a_2e_1 & a_3e_1 \\ 0 & a_2e_2 & a_3e_2 \\ 0 & 0 & a_3e_3 \end{bmatrix} = . \begin{bmatrix} 3 & 1 & \frac{2}{3} \\ 0 & 2 & \frac{2}{3} \\ 0 & 0 & \frac{1}{3} \end{bmatrix}$$

$$A = \begin{bmatrix} 1 & -1 & 0 \\ 2 & 0 & 0 \\ 2 & 2 & 1 \end{bmatrix} = \begin{bmatrix} \frac{1}{3} & -\frac{2}{3} & \frac{2}{3} \\ \frac{2}{3} & -\frac{1}{3} & \frac{-2}{3} \\ \frac{2}{3} & \frac{2}{3} & \frac{1}{3} \end{bmatrix}. \begin{bmatrix} 3 & 1 & \frac{2}{3} \\ 0 & 2 & \frac{2}{3} \\ 0 & 0 & \frac{1}{3} \end{bmatrix} = QR$$

Exercise: Factorize a matrix $A = \begin{bmatrix} 1 & 1 & 2 \\ 0 & 0 & 1 \\ 1 & 0 & 0 \end{bmatrix}$ by the Gram-Schmidt process.

Solution: $A = \begin{bmatrix} 1 & 1 & 2 \\ 0 & 0 & 1 \\ 1 & 0 & 0 \end{bmatrix} = QR = \begin{bmatrix} \frac{1}{\sqrt{2}} & \frac{1}{\sqrt{2}} & 0 \\ 0 & 0 & 1 \\ \frac{1}{\sqrt{2}} & -\frac{1}{\sqrt{2}} & 0 \end{bmatrix} \begin{bmatrix} \sqrt{2} & \frac{1}{\sqrt{2}} & \sqrt{2} \\ 0 & \frac{1}{\sqrt{2}} & \sqrt{2} \\ 0 & 0 & 1 \end{bmatrix}$

2.2 CALCULUS AND OPTIMIZATION TECHNIQUES

Calculus and optimization methods play an important role in data analysis algorithms. Data science algorithms can be viewed as an optimization problem and corresponding numerical methods provide learning rules for the algorithms.

The second section of this chapter begins with basic concepts of calculus like partial derivatives, gradient, directional derivatives, and Jacobian and Hessian matrix.

Constrained and unconstrained optimization techniques, conditions for optimality, gradient descent methods for optimization, and convex functions and their properties are discussed in detail with examples.

- 2.2.1 Introduction to multivariate calculus: partial derivatives, gradient, directional derivatives, Jacobian, Hessian matrix, convex sets and convex functions
- 2.2.2 Constrained and unconstrained optimization techniques

2.2.1 INTRODUCTION TO MULTIVARIATE CALCULUS

Machine learning is about minimizing a cost function that is a scalar function of several variables that typically measures how the model fits into the data and thus calculus plays an important role in understanding the algorithms of machine learning. Multivariable calculus is the extension of calculus in one variable to several variables.

Functions of n variables
Let $S \subseteq R^n$. A real valued function f on S is a rule that assigns to every element $x_1, x_2, ...,x_n$ of S to a variable w given as $w = f(x_1, x_2, ... ,x_n) \; \forall \; (x_1, x_2, ... ,x_n) \in S$, $w \in R^n$.

The variables $x_1, x_2, ...,x_n$ are called independent variables and w is called a dependent variable.

Domain and range of a function of n variables
Consider a function $f: R^n \rightarrow R$ defined as: $f(x_1, x_2, ... ,x_n) = z$

DOMAIN: The set of all points $(x_1, x_2, ... ,x_n) \in R^n$ for which $f(x_1, x_2, ... ,x_n)$ is defined is called the domain of f, denoted as D(f) or Dom(f).

$D(f) = \{(x_1, x_2, ... ,x_n) \in R^n | f(x_1, x_2, ... ,x_n) \text{ is defined}\}$

RANGE: The set of all images of the points $(x_1, x_2, \ldots, x_n) \in D(f)$ of the domain, denoted as $R(f)$ or Range(f).

$$R(f) = \{f(x_1, x_2, \ldots, x_n) | (x_1, x_2, \ldots, x_n) \in D(f)\}$$

A. Partial derivatives of a function:

Consider a function of n variables, $z = f(x_1, x_2, \ldots, x_n)$. Then, the partial derivative of f with respect to an independent variable x_i for any i = 1,2, ...,n, denoted as f_{x_i} or $\frac{\partial f}{\partial x_i}$ and is given as

$$f_{x_i} = \frac{\partial f}{\partial x_i} = \lim_{\Delta x_i \to 0} \frac{f(x_1, x_2, \ldots, x_i + \Delta x_i) - f(x_1, x_2, \ldots, x_n)}{\Delta x_i} \text{ provided the limit exists.}$$

B. Gradient of a function:

The single most important concept from calculus in the context of data science is the gradient. Gradients generalize derivatives to scalar functions of several variables.

The gradient of a function $f: R^n \to R$ denoted by ∇f, is given as

$$\nabla f = \left[\frac{\partial f}{\partial x_1} \quad \frac{\partial f}{\partial x_2} \quad \cdot \ \cdot \ \cdot \quad \frac{\partial f}{\partial x_n} \right]^T$$

i.e., $[\nabla f]_i = \frac{\partial f}{\partial x_i}$

The gradient of function (∇f) points in the direction of steepest ascent from x. Similarly, $-(\nabla f))$ points in the direction of steepest descent from x.

Example. Gradient of the function $f(x, y) = x^2 - xy + y^2$

$$\nabla f = \left[\frac{\partial f}{\partial x} \quad \frac{\partial f}{\partial y} \right]^T = [2x - y \quad -x + 2y]^T$$

C. Directional derivatives of a function:

The rate of change of the function $f(x_1, x_2)$ of two variables in the direction of unit vector $\vec{u} = \langle a_1, a_2 \rangle$ is called the directional derivative of f in the direction of $\vec{u}$ denoted by $D_{\vec{u}} f(x_1, x_2)$.

$$D_{\vec{u}} f(x_1, x_2) = \lim_{h \to 0} \frac{f(x_1 + a_1, x_2 + a_2 h) - f(x_1, x_2)}{h}$$

Example. Directional derivative of the function $f(x, y) = xy^2$ in the direction of (1,2) at the point (3,2).

Solution: $\nabla f(3, 2) = \left[\frac{\partial f}{\partial x} \quad \frac{\partial f}{\partial y} \right](3, 2) = [y^2 \quad 2xy](3, 2) = [4 \quad 12]$

$$u(x, y) = [1 \quad 2] - [3 \quad 2] = [-2 \quad 0]$$

Unit vector in the direction of (1,2) from (3,2) is $\frac{u(x,y)}{|u|} = \frac{[-2\ 0]}{2} = [-1\ 0]$

Directional derivative of $f(x, y) = xy^2$ in the direction of (1,2) at the point (3,2) is $\nabla f(3, 2) \cdot \frac{u(x,y)}{|u|} = [4\ 12] \cdot [-1\ 0]^T = -4$.

D. Jacobian of a function:

The Jacobian of $f: R^n \to R^m$ is a matrix of first-order partial derivatives.

$$J_f = \begin{bmatrix} \frac{\partial f_1}{\partial x_1} & \cdots & \frac{\partial f_1}{\partial x_n} \\ \cdots & \cdots & \cdots \\ \frac{\partial f_m}{\partial x_1} & \cdots & \frac{\partial f_m}{\partial x_n} \end{bmatrix} \text{ i.e., } \left[J_f \right]_{ij} = \frac{\partial f_i}{\partial x_j}$$

Note the special case m = 1, where $\nabla f = J_f^T$.

Example Jacobian matrix of the function $f(x, y, z) = x^2 + 2xy^2 + z^2$, $g(x, y, z) = x^2 + xy$, $h(x, y, z) = y^2 + xz^2$

$$J_f = \begin{bmatrix} \frac{\partial f}{\partial x} & \frac{\partial f}{\partial y} & \frac{\partial f}{\partial z} \\ \frac{\partial g}{\partial x} & \frac{\partial g}{\partial y} & \frac{\partial g}{\partial z} \\ \frac{\partial h}{\partial x} & \frac{\partial h}{\partial y} & \frac{\partial h}{\partial z} \end{bmatrix} = \begin{bmatrix} 2x + 2y^2 & 2xy & 2z \\ 2x + y & x & 0 \\ 0 & 2y & 2xz \end{bmatrix}$$

E. The Hessian matrix of a function:

The Hessian matrix of $f: R^d \to R$ is a matrix of second-order partial derivatives:

$$\nabla^2 f = \begin{bmatrix} \frac{\partial^2 f}{\partial x_1^2} & \cdots & \frac{\partial^2 f}{\partial x_1 \partial x_d} \\ \cdots & \cdots & \cdots \\ \frac{\partial^2 f}{\partial x_d \partial x_1} & \cdots & \frac{\partial^2 f}{\partial x_d^2} \end{bmatrix} \text{ i.e., } \left[J_f \right]_{ij} = \frac{\partial f_i}{\partial x_j}$$

The Hessian matrix is used in some optimization algorithms, such as Newton's method. It is expensive to calculate but can drastically reduce the number of iterations needed to converge to a local minimum by providing information about the curvature of function.

Example Hessian matrix of the function $f(x, y) = x^2 - xy + y^2$

$$\nabla^2 f = \begin{bmatrix} \frac{\partial^2 f}{\partial x^2} & \frac{\partial^2 f}{\partial x \partial y} \\ \frac{\partial^2 f}{\partial x \partial y} & \frac{\partial^2 f}{\partial y^2} \end{bmatrix} = \begin{bmatrix} 2 & -1 \\ -1 & 2 \end{bmatrix}$$

F. Convexity:

While dealing with the algorithms related to machine learning, "convexity" plays a vital role. Several results have been developed in optimization theory based on the

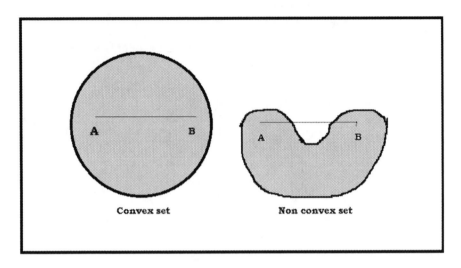

FIGURE 2.1 Convex and Nonconvex Set.

concept of convexity. *Convexity* is a term that pertains to both sets and functions. For functions, there are different degrees of convexity, and how convex a function is tells us a lot about its minima: do they exist, are they unique, how quickly we can find them using optimization algorithms, etc.

Convex sets:
A set $S \subseteq R^d$ is convex if $tx + (1 - t)y \in S$ for all $x, y \in S$ and $t \in [0, 1]$.

Geometrically, this means that all the points on the line segment between any two points in χ are also in χ

Convex functions:
Let $f: R^d \to R$ A function f is convex if $f(tx + (1 - t)y) \leq tf(x) + (1 - t)f(y)$ for all $x, y \in dom(f)$ and $t \in [0, 1]$ is represented in Figures 2.1

2.2.2 CONSTRAINED AND UNCONSTRAINED OPTIMIZATION TECHNIQUES

The optimization technique is the method of finding a set of inputs to an objective function that results in a maximum or minimum function evaluation. It is the challenging problem that underlies many machine learning algorithms, from fitting logistic regression models to training artificial neural networks. When defining extrema, it is necessary to consider the set of inputs over which we're optimizing. This set X is a subset of R^d and is called the feasible set. If X is the entire domain of the function being optimized, then the problem is unconstrained; otherwise, the problem is constrained.

Unconstrained optimization is the process of optimizing an objective function with respect to some variables with no constraints on it. Some of the methods used to solve unconstrained optimization problem are Newton's method and the gradient descent method.

Newton's Method

Example Optimize the function $f = x^2 - 2xy + 4y^2$ from the point $a = \begin{bmatrix} -3 \\ 1 \end{bmatrix}$.

At the point $a = \begin{bmatrix} -3 \\ 1 \end{bmatrix}$ $\nabla f = \begin{bmatrix} -8 \\ 14 \end{bmatrix}$ and Hessian matrix $H = \begin{bmatrix} 2 & -2 \\ -2 & 8 \end{bmatrix}$.

The Hessian inverse matrix is $H^{-1} = \begin{bmatrix} 0.6667 & 0.16667 \\ 0.16667 & 0.16667 \end{bmatrix}$.

Thus, $\Delta x = -\begin{bmatrix} 0.6667 & 0.16667 \\ 0.16667 & 0.16667 \end{bmatrix}\begin{bmatrix} -8 \\ 14 \end{bmatrix} = \begin{bmatrix} 3 \\ -1 \end{bmatrix}$.

So $x_1 = x + \Delta x = \begin{bmatrix} -3 \\ 1 \end{bmatrix} + \begin{bmatrix} 3 \\ -1 \end{bmatrix} = \begin{bmatrix} 0 \\ 0 \end{bmatrix}$.

Constrained optimization is the process of optimizing an objective function with respect to some variables in the presence of constraints on it.

Some of the methods used to solve constrained optimization problem are Lagrange's multiplier method, linear programming, and quadratic programming.

Example 1: Find the minimum value of $x^2 + y^2$ subject to the condition $ax + by = c$ using Lagrange's multiplier method.

Solution: Let

$$u = f(x, y) = x^2 + y^2 \tag{2.1}$$

$$\text{condition } \emptyset = ax + by - c = 0 \tag{2.2}$$

Construct the function $F = u + \lambda\emptyset = (x^2 + y^2) + \lambda(ax + by - c)$.

To find a stationary point, form the equations

$$\frac{\partial F}{\partial x} = 0 \; 2x + a\lambda = 0 \tag{2.3}$$

$$\frac{\partial F}{\partial y} = 0 \; 2y + b\lambda = 0 \tag{2.4}$$

We eliminate x, y and λ using the equations (2.1), (2.2), (2.3), (2.4).
From equations 2.3 and 2.4:

$$\lambda = \frac{2x}{a}, \lambda = \frac{2y}{b} \rightarrow \frac{2x}{a} = \frac{2y}{b}$$

$$\frac{2x}{a} = \frac{2y}{b} = k \ (say)$$

From equations (2.2) and (2.5):

$$ax + by - c = 0 \ implies \frac{a^2k}{2} + \frac{b^2k}{2} = c$$

$$k = \frac{2c}{a^2 + b^2}$$

$$\therefore x = \frac{a}{2}\left[\frac{2c}{a^2 + b^2}\right], \ y = \frac{b}{2}\left[\frac{2c}{a^2 + b^2}\right]$$

Substituting these values in (1):

$$u = x^2 + y^2 = \frac{a^2}{4}\left[\frac{2c}{a^2 + b^2}\right]^2 + \frac{b^2}{4}\left[\frac{2c}{a^2 + b^2}\right]^2$$

$$= \frac{c^2(a^2 + b^2)}{(a^2 + b^2)^2}$$

$$= \frac{c^2}{a^2 + b^2}$$

Exercise: Use the Lagrange method to find the minimum distance from origin to the plane $3x + 2y + z = 12$ [**Solution:** minimum distance from origin to the plane is $\sqrt{\frac{504}{7}}$.]

2.3   REGRESSION ANALYSIS

INTRODUCTION

Regression analysis is widely used for prediction and forecasting. Regression in data science consists of mathematical methods that allow data scientists to predict a continuous outcome (y) based on the value of one or more predictor variables (x). This statistical method provides relation between two correlated data variables. It is usually used to estimate values of one variable corresponding to the specified values of other variable. If the variables are not strongly correlated, then it is possible to draw several lines through given data set. Based on the method of least squares, a best fit line can be obtained. Linear regression and logistic regression are probably the most prominent techniques used in data science related to civil engineering problems.

2.3.1 Simple Linear Regression
2.3.2 Multiple Linear Regression
2.3.3 Polynomial Regression Analysis

2.3.4 Logistic Regression Analysis
2.3.5 Lasso Regression

2.3.1 Simple Linear Regression

Linear regression quantifies the relationship between one or more predictor variable (s) and one outcome variable. Simple linear regression is one of the simplest and most powerful regression techniques. It helps us to predict output (y) from the trained input values (x) by fitting a straight line between those variables.

A best filt line is represented mathematically as y = a x+ b, where 'x' and 'y' input and output variables.

$$a = \frac{n\sum_{i=1}^{n} x_i y_i - \sum_{i=1}^{n} x_i \sum_{i=1}^{n} y_i}{n\sum_{i=1}^{n} x_i^2 - \left(\sum_{i=1}^{n} x_i\right)^2}$$

and

$$b = \frac{1}{n}\left(\sum_{i=1}^{n} y_i - a \sum_{i=1}^{n} x_i\right)$$

Where 'a' and 'b' are calculated using the least squares approach by minimizing the sum of squares of deviation $\min_{w} \|X_w - y\|_2^2$ are represented in Figures 2.2 and 2.3

Examples on linear regression:

Example 1: The sales building blocks of a construction company (in million dollars) for each year are shown in the table below.

x(year)	2005	2006	2007	2008	2009
y(sales)	12	19	29	37	45

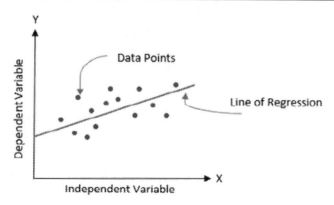

FIGURE 2.2 Regression Line.

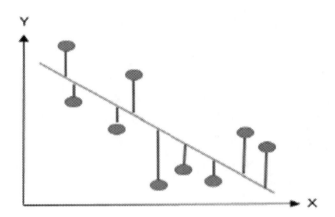

FIGURE 2.3   Regression Line.

Find the linear regression line y = ax + b. Also estimate the sales of the company in 2012.

Solution:
We first change the variable x into t such that t = x – 2005 and therefore t represents the number of years after 2005. Using t instead of x makes the numbers smaller and therefore manageable. The table of values becomes.

t	y	ty	t^2
0	12	0	0
1	19	19	1
2	29	58	4
3	34	111	9
4	45	180	16
$\Sigma x = 10$	$\Sigma y = 142$	$\Sigma xy = 368$	$\Sigma x^2 = 30$

We now calculate a and b using least squares regression formulas:

$$a = \frac{n\sum_{i=1}^{n} x_i y_i - \sum_{i=1}^{n} x_i \sum_{i=1}^{n} y_i}{n\sum_{i=1}^{n} x_i^2 - \left(\sum_{i=1}^{n} x_i\right)^2} = 8.4$$

and

$$b = \frac{1}{n}\left(\sum_{i=1}^{n} y_i - a \sum_{i=1}^{n} x_i\right) = 11.6$$

Therefore, y = 8.4x + 11.6.

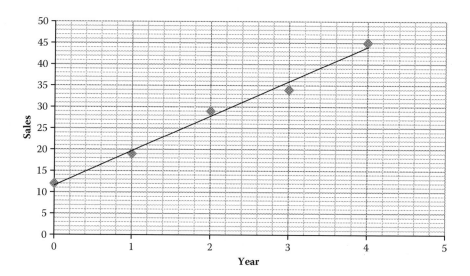

FIGURE 2.4   Regression Line.

In 2012, t = 2012 − 2005 = 7.

The estimated sales in 2012 are: y = 8.4 × 7 + 11.6 = 70.4 million dollars. Its decision boundary is represented in Figure 2.4

Example 2: For the following set of data, find the linear regression line.

x	0	1	2	3	4
y	2	3	5	4	6

Solution:

x	0	1	2	3	4	$\Sigma x = 10$
y	2	3	5	4	6	$\Sigma y = 20$
xy	0	3	10	12	24	$\Sigma xy = 49$
x^2	0	1	4	9	16	$\Sigma x^2 = 30$

We now calculate a and b using the least square regression formulas:

$$a = \frac{n \sum_{i=1}^{n} x_i y_i - \sum_{i=1}^{n} x_i \sum_{i=1}^{n} y_i}{n \sum_{i=1}^{n} x_i^2 - \left(\sum_{i=1}^{n} x_i \right)^2} = 0.9$$

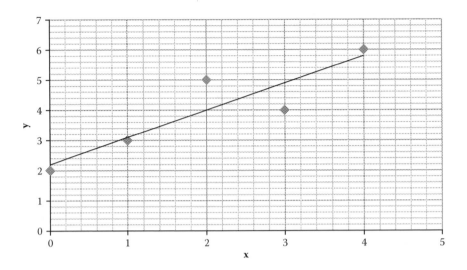

FIGURE 2.5   Regression Line.

and

$$b = \frac{1}{n}\left(\sum_{i=1}^{n} y_i - a \sum_{i=1}^{n} x_i\right) = 2.2$$

Therefore, $y = 0.9x + 2.2$.

Substituting x by 10 we get corresponding value of y, we get $y = 0.9 \times 10 + 2.2 = 11.2$ and it is represented in Figure 2.5

Exercise:

The given data set is the salary data set. It contains two columns: year of experience and salary of different employees. Create a model using Python that predicts the salary of the employee based on the years of experience. Since it contains one independent and one dependent variable, we can use simple linear regression for this problem.

Years of Experience (X)	1.1	1.3	1.5	2	2.2	2.9	3	3.2	3.3
Salary (Y)	39343	46205	37731	43525	39891	56642	60150	54445	64445

2.3.2   MULTIPLE LINEAR REGRESSION

Multivariate linear regression deals with multiple independent variables. This is similar to simple linear regression, but there is more than one independent variable. Every value of the independent variable 'x' is associated with a value of the dependent variable y.

Mathematically, it's expressed by:

$$y = b_0 + b_1x_1 + b_2x_2 + b_3x_3 + \dots$$

In the case of two independent variables, the equation reduces to that of a plane, $y = b_0 + b_1X_1 + b_2X_2$:

Where,

$$b_0 = \bar{Y} - b_1\bar{X_1} - b_2\bar{X_2}$$

$$b_1 = \frac{\sum x_2^2 \sum x_1 y - \sum x_1 x_2 \sum x_2 y}{\sum x_1^2 \sum x_2^2 - (\sum x_1 x_2)^2}$$

$$b_2 = \frac{\sum x_1^2 \sum x_2 y - \sum x_1 x_2 \sum x_1 y}{\sum x_1^2 \sum x_2^2 - (\sum x_1 x_2)^2}$$

Where,

$$\sum x_1^2 = \sum X_1 X_1 - \frac{\sum X_1 \sum X_1}{N}, \quad \sum x_2^2 = \sum X_2 X_2 - \frac{\sum X_2 \sum X_2}{N}$$

$$\sum x_1 y = \sum X_1 Y - \frac{\sum X_1 \sum Y}{N}, \quad \sum x_2 y = \sum X_2 Y - \frac{\sum X_2 \sum Y}{N}$$

$$\sum x_2 x_2 = \sum X_1 X_2 - \frac{\sum X_1 \sum X_2}{N}$$

Example 1: Find the regression line for the following set of data. Also, estimate the value of Y for $x_1 = 3$ and $x_2 = 2$.

Subject	Y	X_1	X_2	X_1X_1	X_2X_2	X_1X_2	X_1Y	X_2Y
1	−3.7	3	8	9	64	24	−11.1	−29.6
2	3.5	4	5	16	25	20	14	17.5
3	2.5	5	7	25	49	35	12.5	17.5
4	11.5	6	3	36	9	18	69	34.5
5	5.7	2	1	4	1	2	11.4	5.7
Σ	19.5	20	24	90	148	99	95.8	45.6

$$\sum x_1^2 = \sum X_1 X_1 - \frac{\sum X_1 \sum X_1}{N} = 10, \quad \sum x_2^2 = \sum X_2 X_2 - \frac{\sum X_2 \sum X_2}{N} = 32.8$$

$$\sum x_1 y = \sum X_1 Y - \frac{\sum X_1 \sum Y}{N} = 17.8, \quad \sum x_2 y = \sum X_2 Y - \frac{\sum X_2 \sum Y}{N} = -48$$

$$\sum x_2 x_2 = \sum X_1 X_2 - \frac{\sum X_1 \sum X_2}{N} = 3, \quad b_1 = \frac{\sum x_2^2 \sum x_1 y - \sum x_1 x_2 \sum x_2 y}{\sum x_1^2 \sum x_2^2 - (\sum x_1 x_2)^2} = 2.28$$

$$b_2 = \frac{\sum x_1^2 \sum x_2 y - \sum x_1 x_2 \sum x_1 y}{\sum x_1^2 \sum x_2^2 - (\sum x_1 x_2)^2} = -1.67$$

$$b_0 = \bar{Y} - b_1 \bar{X_1} - b_2 \bar{X_2} = \frac{19.5}{5} - \frac{2.28 * 20}{5} - \frac{-1.67 * 24}{5} = 2.796$$

Therefore, the regression equation becomes

$$Y = 2.796 + 2.28 x_1 - 1.67 x_2$$

We can find Y for given $x_1 = 3$ and $x_2 = 2$:

$$Y = 2.796 + 2.28 * 3 - 1.67 * 2 = 6.296$$

2.3.3 POLYNOMIAL REGRESSION

This method is used to find a polynomial function that gives the best fit to a given sample of data. A linear regression model helps to understand patterns for a given data set by fitting in a simple linear equation, but it might not be accurate while dealing with complex data. In those instances, we need to come up with curves that adjust with the data rather than the lines. One approach is to use a polynomial model whose degree is greater than 1.

Mathematically, a polynomial model is expressed by:

$$y = b_0 + b_1 x + b_2 x^2 + b_3 x^3 + \dots b_n x^n,$$

where 'y' is the predicted value for the polynomial model with regression coefficients b_1 to b_n for each degree and a bias of b_0.

Method

We will develop techniques that fit linear, quadratic, cubic, quartic, and quintic regressions. In fact, this technique will work for any order polynomial.

$$f(x) = b_0 + b_1 x + b_2 x^2 + b_3 x^3 + \dots b_n x^n \quad \text{Where } 1 \le n \le 5$$

The function will only approximate the value that we have in our sample data and consequently there will be a residual error.

$$y_i = f(x_i) + e_i$$

Let us consider the set of sample observations (x_i, y_i).

For each observation i, we will have an equation:

$$y_1 = b_0 + b_1 x_1 + b_2 x_1^2 + b_3 x_1^3 + \dots b_n x_1^n$$
$$y_2 = b_0 + b_1 x_2 + b_2 x_2^2 + b_3 x_2^3 + \dots b_n x_2^n$$
$$\vdots$$
$$y_i = b_0 + b_1 x_i + b_2 x_i^2 + b_3 x_i^3 + \dots b_n x_i^n$$

$$
\begin{bmatrix}
1 & x_1 & x_1^2 & \dots & x_1^n \\
1 & x_2 & x_2^2 & \dots & x_2^n \\
1 & x_3 & x_3^2 & \dots & x_3^n \\
\dots & \dots & \dots & \dots & \dots \\
1 & x_i & x_i^2 & \dots & x_i^n
\end{bmatrix}
\begin{bmatrix}
b_0 \\
b_1 \\
b_2 \\
\dots \\
b_i
\end{bmatrix}
=
\begin{bmatrix}
y_1 \\
y_2 \\
y_3 \\
\dots \\
y_i
\end{bmatrix}
$$

Or, more conveniently,

$$\hat{y} = M\hat{\alpha}$$

Multiply the equation by the transpose of matrix 'M', which is denoted by M'.

We get, $M'\hat{y} = M'M\hat{\alpha}$.

Pre-multiplying by $(M'M)^{-1}$.

$(M'M)^{-1}(M'\hat{y}) = (M'M)^{-1}(M'M)\hat{\alpha}$, We get $\hat{\alpha} = (M'M)^{-1}(M'\hat{y})$

Where,

$$
M'M =
\begin{bmatrix}
1 & 1 & 1 & \dots & 1 \\
x_1 & x_2 & x_3 & \dots & x_i \\
x_1^2 & x_2^2 & x_3^2 & \dots & x_i^2 \\
\dots & \dots & \dots & \dots & \dots \\
x_1^n & x_2^n & x_3^n & \dots & x_i^n
\end{bmatrix}
\begin{bmatrix}
1 & x_1 & x_1^2 & \dots & x_1^n \\
1 & x_2 & x_2^2 & \dots & x_2^n \\
1 & x_3 & x_3^2 & \dots & x_3^n \\
\dots & \dots & \dots & \dots & \dots \\
1 & x_i & x_i^2 & \dots & x_i^n
\end{bmatrix}
$$

$$
M'M =
\begin{bmatrix}
I & \Sigma x_i & \Sigma x_i^2 & \dots & \Sigma x_i^n \\
\Sigma x_i & \Sigma x_i^2 & \Sigma x_i^3 & \dots & \Sigma x_i^{n+1} \\
\Sigma x_i^2 & \Sigma x_i^3 & \Sigma x_i^4 & \dots & \Sigma x_i^{n+2} \\
\dots & \dots & \dots & & \dots \\
\Sigma x_i^n & \Sigma x_i^{n+1} & \Sigma x_i^{n+2} & \dots & \Sigma x_i^{2n}
\end{bmatrix}
$$

$$
M'\hat{y} =
\begin{bmatrix}
\Sigma y_i \\
\Sigma x_i y_i \\
\Sigma x_i^2 y_i \\
\dots \\
\Sigma x_i^n y_i
\end{bmatrix}
$$

Thus, the matrix representation of polynomial regression for any degree polynomial is given by

$$\begin{bmatrix} I & \Sigma x_i & \Sigma x_i^2 & \cdots & \Sigma x_i^n \\ \Sigma x_i & \Sigma x_i^2 & \Sigma x_i^3 & \cdots & \Sigma x_i^{n+1} \\ \Sigma x_i^2 & \Sigma x_i^3 & \Sigma x_i^4 & \cdots & \Sigma x_i^{n+2} \\ \cdots & \cdots & \cdots & \cdots & \cdots \\ \Sigma x_i^n & \Sigma x_i^{n+1} & \Sigma x_i^{n+2} & \cdots & \Sigma x_i^{2n} \end{bmatrix} \begin{bmatrix} b_0 \\ b_1 \\ b_2 \\ \cdots \\ b_m \end{bmatrix} = \begin{bmatrix} \Sigma y_i \\ \Sigma x_i y_i \\ \Sigma x_i^2 y_i \\ \cdots \\ \Sigma x_i^n y_i \end{bmatrix}$$

Where m is the degree of the polynomial and n is the number of data points.

Polynomial regression is used when the data is non-linear. In this, the model is more flexible as it plots a curve between the data. The degree of polynomial needs to vary such that overfitting doesn't occur. Using polynomial regression, we can fit curved lines flexibly between the data, but sometimes even these result in false predictions as they fail to interpret the input due to over fitting. Using regularization, we improve the fit so that accuracy is better on the test data set. Regularization tends to avoid overfitting by adding a penalty term to the cost/loss function.

Example: Consider five observations and find the best fit.

$$x = \{1, 2, 3.5, 5, 6.2\}$$
$$y = \{-7.5, -15.6, -40.5, -80.7, -123.876\}$$

Solution:

$$x = \{1, 2, 3.5, 5, 6.2\} \Rightarrow \Sigma x = 17.7$$
$$y = \{-7.5, -15.6, -40.5, -80.7, -123.876\} \Rightarrow \Sigma y = 268.176$$
$$x^2 = \{1, 4, 12.25, 25, 38.44\} \Rightarrow \Sigma x^2 = 80.69$$
$$xy = \{-7.5, -31.2, -141.75, -403.5, -768.0312\} \Rightarrow \Sigma xy = -1351.981$$

$$\hat{\alpha} = (M'M)^{-1}(M'\hat{y})$$

$$M'M = \begin{bmatrix} i & \Sigma x_i \\ \Sigma x_i & \Sigma x_i^2 \end{bmatrix} = \begin{bmatrix} 5 & 17.7 \\ 17.7 & 80.69 \end{bmatrix}$$

$$M'\hat{y} = \begin{bmatrix} \Sigma y_i \\ \Sigma x_i y_i \end{bmatrix} = \begin{bmatrix} 268.176 \\ -1351.981 \end{bmatrix}$$

$$\hat{\alpha} = \begin{bmatrix} 5 & 17.7 \\ 17.7 & 80.69 \end{bmatrix}^{-1} \begin{bmatrix} 268.176 \\ -1351.981 \end{bmatrix} = \begin{bmatrix} 25.40974 \\ -22.32908 \end{bmatrix}$$

Hence, y = 25.40974 − 22.32908x would be a good linear regression for the data.

However, if we repeat the analysis again, but we try to fit a quadratic regression we get this:

$$M'M = \begin{bmatrix} i & \Sigma x_i & \Sigma x_i^2 \\ \Sigma x_i & \Sigma x_i^2 & \Sigma x_i^3 \\ \Sigma x_i^2 & \Sigma x_i^3 & \Sigma x_i^4 \end{bmatrix} = \begin{bmatrix} 5 & 17.7 & 80.69 \\ 17.7 & 80.69 & 415.203 \\ 80.69 & 415.203 & 2269.696 \end{bmatrix}$$

$$M'\hat{y} = \begin{bmatrix} \Sigma y_i \\ \Sigma x_i y_i \\ \Sigma x_i^2 y_i \end{bmatrix} = \begin{bmatrix} 268.176 \\ -1351.981 \\ -7345.318 \end{bmatrix}$$

$$\hat{\alpha} = \begin{bmatrix} 5 & 17.7 & 80.69 \\ 17.7 & 80.69 & 415.203 \\ 80.69 & 415.203 & 2269.696 \end{bmatrix}^{-1} \begin{bmatrix} 268.176 \\ -1351.981 \\ -7345.318 \end{bmatrix} = \begin{bmatrix} -6.2 \\ 2.1 \\ -3.4 \end{bmatrix}$$

Hence, $y = -3.4x^2 + 2.1x - 6.2$.

2.3.4 LOGISTIC REGRESSION

Logistic regression is used to develop a model that relates categorical variables with predictor variables. It is generally used as a classification algorithm when categorical variables are of the type Yes/No, True/False, 0/1, etc. This can be classified into a binary that has two classes and multi-class logistic regression with more than two classes.

Logistic regression is capable of finding out the probability only after transforming the dependent variable into a logit variable with respect to the independent variable or the features present in the data.

A logit function is defined as a linear function of variables 'Xi' and is written as $Log(P/Q) = \mu_0 + \mu_1 x_1 + \mu_2 x_2 + \ldots + \mu_r x_r$, where 'P' represents probability of success of an occurrence of an event and 'Q=1−P' represents the probability of an event not occurring.

The corresponding logistic regression model is computed as

$$P(X) = \frac{1}{1 + \exp(-\mu_0 + \mu_1 x_1 + \mu_2 x_2 + \ldots + \mu_r x_r)}$$

Since we are working here with a binomial distribution (dependent variable), the best-suited link function is selected based on the maximum likelihood of samples that are considered. It is mathematically represented as a sigmoid function that gives probability lying between 0 to 1 and it is represented in Figure 2.6

In logistic regression, we generally compute the probability that lies between the interval 0 and 1 (inclusive of both), which can be used to classify the data. Logistic regression is used to classify the data whereas linear regression is used for generating continuous values. Logistic regression works fine only when the target variable is discrete in nature. Moreover, linear regression does not provide model for classification problems as the predicted value may exceed the range (0,1) or error rate increases if the

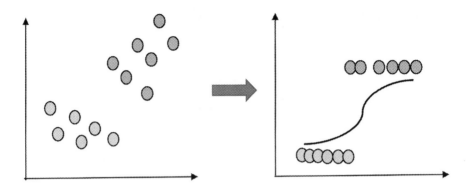

FIGURE 2.6   Sigmoid Representation for Predicted and Targeted Variables.

data has outliers. For logistic regression, the data size should be large and non-collinearity must be observed among the feature variables. There is less chance of overfitting and logistic can overfit in case of higher dimension. For such cases, regularizing techniques that shrink the coefficients may be used to avoid overfitting.

2.3.5   LASSO REGRESSION

LASSO regression is one of the simple techniques to reduce model complexity and prevent over-fitting. This is one of widely used techniques that is aimed at decreasing the size of the coefficients by penalizing the sum of absolute values of the coefficients.

LASSO is a short form of least absolute shrinkage and selection operator. This methods provides better accuracy as compared to other regression methods. In this method, data values are shrunk in the direction of the mean. This method is especially useful when multi-collinearity exhibits among data values or eliminates certain parameters or variables in the subparts of the model.

LASSO regression is expressed mathematically as $\sum_{i=1}^{n}\left(y_i - \sum_{j} x_{ij}\gamma_j\right)^2 + \alpha\sum_{j=1}^{p}|\gamma_j|$.

Where 'α' represents the amount of shrinkage, the first term represents sum of squares of residuals, and the second term represents the sum of modulus value of the coefficients. This is useful especially for predicting cost of huge construction management projects.

Different types of regression analysis techniques can be used to build the model, depending upon the kind of data available or the one that gives the maximum accuracy.

REFERENCE

1. David C. Lay, Steven R. Lay, Judi J. McDonald. "Linear Algebra and Its Applications", Fifth Edition, 2016, Pearson.

3 Data Analytics for Environmental Engineering

Ravindra K. Lad and R. A. Christian

3.1 INTRODUCTION TO ENVIRONMENTAL ENGINEERING

The main areas of environmental engineering include air pollution control, noise pollution control, water supply, wastewater management, solid waste management, hazardous waste management, biomedical waste management, and land management.

In environmental engineering, to remediate pollution, the stages required are collection of data, analysis of data, design, and treatment.

3.1.1 ROLE OF DATA ANALYTICS IN ENVIRONMENTAL ENGINEERING

This field involves the study of different environmental sectors like air, noise, water, wastewater, and solid waste. Predicting and estimating parameters of these sectors for resolving the impact on environment of these parameters are challenging for environmental engineers. As environmental engineering data is rising in complexity, resolution, and size, big data is required for appropriate treatment or resolving the pollution problems. This big data supports complete analysis and transforms this into knowledge that requires the application of data science.

We have perceived in recent eras that the popularity of utilization of data science methods is increasing in different domains of organizations, including in the area of environmental engineering also.

After collection of data, normally the mean of data is considered for further analysis. An arithmetic mean cannot be used for all types of data. The consideration of type of mean depends on the type of data. For example, in the case of wastewater analysis, the geometric mean is required as there is a possibility of vast variations in data. The examples of choosing the appropriate measure of central tendency related to environmental engineering data are required.

In environmental engineering data analysis, sample size is the most important parameter for the precision of our estimates and the power of the study to draw conclusions. The article 3.2.2 will cover the determination of sample size with example.

DOI: 10.1201/9781003316671-3

3.2   DATA ANALYSIS IN ENVIRONMENTAL ENGINEERING

3.2.1   TYPES OF SAMPLE COLLECTION

The sampling techniques used in a wastewater or water survey must assume that a representative sample is obtained. There are two methods of sampling: one is grab and another is composite sampling. These two methods are explained below.

3.2.1.1   Grab Sampling

In the case of grab sampling, a sample is required to be collected at a point where sewage/wastewater/water can be mixed thoroughly.

3.2.1.2   Composite Sampling

In the case of composite sampling, the sample is collected at regular intervals during a day. These different samples are now mixed together and utilized from each specimen and are proportional to the rate of flow at the time the specimen was collected.

3.2.2   DETERMINATION OF SAMPLE SIZE

3.2.2.1   Confidence Interval and Confidence Limits

As there is a possibility of variability in estimates in other samples, it is better to avoid a sample point estimate. To specify this variability, interval estimates are used. A shortcoming of point estimates, like a mean bio-chemical oxygen demand (BOD) of wastewater sample is 15 mg/L, from this value we cannot predict the status of BOD. But the status of a parameter can be predicted from interval estimates; for example, the interval from say 10 to 20 mg/L. Interval estimates are also known as confidence intervals. The confidence intervals define two limits: one is an upper limit (20 mg/L) and other is a lower limit (10 mg/L). The ends of the confidence interval are known as the confidence limits. The confidence interval can be developed for any population parameter. The confidence intervals are commonly used for the mean, correlations, and relative risk.

3.2.2.2   Confidence Interval for Means
1. Large Samples (N >30)

When a number of samples is more than 30, then samples are considered large samples. The selection of type of test statistic is the subject of the standard deviation (σ) of the population is known or not known to test a hypothesis for the mean.

   If σ is known, then the Z-statistic is used and if σ is unknown, then the t-statistic is used.

   In case of standard deviation, if (σ) is known, confidence intervals are given as follows:

$$\overline{X} - Z_{\alpha/2}\left(\frac{\sigma}{\sqrt{N}}\right) \le \mu \le \overline{X} + Z_{\alpha/2}\left(\frac{\sigma}{\sqrt{N}}\right), \text{Interval: Two-sided} \qquad (3.1)$$

$$\overline{X} - Z_\alpha\left(\frac{\sigma}{\sqrt{N}}\right) \leq \mu \leq \infty, \text{ Interval: Lower one-sided} \qquad (3.2)$$

$$-\infty \leq \mu \leq \overline{X} + Z_\alpha\left(\frac{\sigma}{\sqrt{N}}\right), \text{ Interval: Upper one-sided} \qquad (3.3)$$

Where,

$\overline{X}$ = mean of the sample;
N = size of the sample;
Z_α and $Z_{\alpha/2}$ = values of random variables (in the case of standard normal distribution) and cutting off $(1 - \lambda)$ or $(1 - \lambda/2)$ percent in the tail of the distribution, respectively.
$\langle$ = level of significance and equals to $1 - \lambda$.

2. Small Samples (N < 30)

When the number of samples is less than 30, then samples are considered small samples. In such cases, equations (3.1 to 3.3) (6.1 to 6.3) cannot be used for the determination of the confidence limits. The real distribution is given by the student t-distribution for small samples. In student t-distribution, the critical value is denoted by $t_{\alpha/2}$. For Z normal statistics, the t-statistic is given as follows:

$$t = \frac{\overline{X} - \mu}{s/\sqrt{N}} \qquad (3.4)$$

The equation 3.5 equation 6.4 yields for the confidence limits:

$$\overline{X} - t_{\alpha/2}\left(\frac{s}{\sqrt{N}}\right) \leq \mu \leq \overline{X} + t_{\alpha/2}\left(\frac{s}{\sqrt{N}}\right), \text{ Interval: Two-sided} \qquad (3.5)$$

$$\overline{X} - t_\alpha\left(\frac{s}{\sqrt{N}}\right) \leq \mu \leq \infty, \text{ Interval: Lower one-sided} \qquad (3.6)$$

$$-\infty \leq \mu \leq \overline{X} + t_\alpha\left(\frac{s}{\sqrt{N}}\right), \text{ Interval: Upper one-sided} \qquad (3.7)$$

Where,

$\overline{X}$ = the observed mean
s = the sample standard deviation

t_α = the confidence coefficient (critical value from t-distribution with
$\upsilon = N-1$).

$\langle$ = the level of significance for one-sided confidence interval
$\alpha/2$ is used for a two-sided confidence interval

In this case, σ is unknown and hence σ is replaced by s.
(Spiegel and Stephens, 1999; Ayyub, 2003 and 2014; Alfassi et al., 2004).

3.2.2.3 Sample Size
The sample size (N) can be calculated as follows. When the population variance is
unknown, equation (6.5) can be used to solve for the half-width of the confidence
interval (W) as

$$W = t_{\alpha/2}\left(\frac{s}{\sqrt{N}}\right)$$
(3.8)

Equation (3.8) can be written as

$$\frac{W}{s} = \frac{t_{\alpha/2}}{\sqrt{N}}$$
(3.9)

The normalized half-width W_N is then

$$W_N = \frac{W}{s}$$
(3.10)

Therefore, the sample size N is

$$N = \left(\frac{t_{\alpha/2}}{W_N}\right)^2 = \left(\frac{t_{\alpha/2}s}{W}\right)^2$$
(3.11)

In order to look up a value for t_α, a sample size is needed. Therefore, the solution
procedure is iterative by assuming a sample size, then looking up the table value of
t_α, then computing the sample size (using equation 3.11) and then repeating the
process.

3.2.2.4 Illustrative Example of Determination of Sample Size
Illustrative example is given below for the determination of sample size.
For example, the standard deviation of BOD of waste water samples of industry
is 1.21.
The number of samples collected is 20.
The statistically acceptable sample size is calculated as under:

$$s = 1.21 \text{ and } N = 20.$$

Assume 90% confidence (i.e., p = 90).

From the t-distribution table, t_α for degree of freedom, $\upsilon = N - 1 = 20 - 1 = 19$ is 1.73 (Spiegel and Stephens (1999).

Now, the half-width of the confidence interval:

$$W = t_{\alpha/2}\left(\frac{s}{\sqrt{N}}\right)$$

$$W = 1.73\left(\frac{1.21}{\sqrt{20}}\right) = 0.468$$

Assuming the standard deviation of BOD of wastewater samples remaining true at 5% level of significance (i.e., p = 95). The sample size is calculated as follows:

From the t-distribution table, t_α for degree of freedom, $\upsilon = N - 1 = 20 - 1 = 19$ is 2.09.

$$N = \left(\frac{t_{\alpha/2}s}{W}\right)^2$$

Therefore,

$$N_1 = \left(\frac{2.09 \times 1.21}{0.468}\right)^2 = 29.19$$

Proceeding in the similar way, looking up the t-distribution table, t_α for degree of freedom, $\upsilon = N - 1 = 29 - 1 = 28$ is 2.05. Then the second sample size is computed as

$$N_2 = \left(\frac{2.05 \times 1.21}{0.468}\right)^2 = 28.$$

Continuing in a similar way, the further result is as follows:

t_α for degree of freedom, $\upsilon = N - 1 = 28 - 1 = 27$ is 2.05. Then,

$$N_3 = 28.$$

As $N_2 = N_3$, the solution is found: 28 samples and the conclusion is that the effective sample size for the above problem is accepted as 28. So the minimum number of samples required is 28. This way, a sample size can be calculated for all parameters of all industries and then data can be generated as per the sample size using the bootstrap technique of MATLAB software. This generated data can be used for further work.

3.3 APPLICATIONS OF SOFT COMPUTING TOOLS

The engineering field is most creative. Hence, problems occur in the field that are more uncertain and unstructured. If we try to find out the solution for such problems by a conventional method, it become time consuming and labour intensive. To overcome this problem, nowadays, soft computing techniques are used, which consist of various techniques and methodology. Soft computing is nothing but an algorithm that is based on the individual's knowledge and skills (Sattar, 2009). Soft computing techniques are used to solve very complex and wide range problems with the advent of low cost and high performance. Soft computing techniques are multi-disciplinary fields. Zadeh stated that the soft computing tool is not only one method, but the soft computing tool is a combination of numerous methods, like fuzzy logic, neural networks, genetic algorithms, etc. (Yousef Sajed et al., 2019). All of these methods can be used individually or in combination as per the type of problem (Buckley and Hayashi, 1994).

3.3.1 ARTIFICIAL NEURAL NETWORK

The concept of artificial neural networks (ANN) was invented by Frank Rosenblatt. The pioneers of work related to artificial neural networks, Warren McCulloch and Walter Pitts (1943) developed a computational model for neural networks, which was based on algorithms i.e., threshold logic. McCulloch and Walter Pitts (1943) paved the way through this model for research. This model split into two approaches for research. One approach stressed on biological processes and other stressed on the application related to artificial intelligence.

Applications of Artificial Neural Networks:

1. For data processing, sequence and pattern recognition systems, robotics, etc.
2. ANN can solve multi aspect issues which are difficult to accomplish.
3. The main application of ANNs is to develop a model to accomplish several tasks of computations faster than the traditional systems.
4. These several tasks include optimization, classification, approximation, and data clustering.

Basically, it is a technique that is effectively used for predictive modeling and analysis, enabling data analysis using a set of operations based on a biological modeling concept resembling human brain and neural functioning.

3.3.2 GENETIC ALGORITHM

John Holland invented the concept of genetic algorithms (GAs). GAs are usually used to generate high-quality solutions to optimization and search problems depend on biologically inspired operators like crossover, mutation, and selection (Emery A. Coppola Jr, et al., 2002). Both kinds of constrained and unconstrained optimization problems can be solved by genetic algorithms that are established on a natural selection process.

Applications of GAs:

1. Genetic algorithms are usually used to generate high-quality solutions to optimization and search problems depend on biologically inspired operators like crossover mutation and selection.
2. Genetic algorithms are basically an optimization method that is based on the natural genetics and selection mechanics.
3. Genetic algorithms follow this principle of natural genetics and natural selection to create optimization and search procedures.
4. GAs are the best tool to get optimum solutions in a short time.

Basically, it is a technique that is effectively used for optimization of variables and deals with a variety of behaviours of variables like linear, non-linear, stochastic, etc.

3.3.3 Fuzzy Logic

The crisp set theory i.e., two-valued logic theory, is insufficient for dealing with uncertainty, imprecision, and complexity of the real world. The uncertainty has a vital role to maximize the usefulness of systems models. Lotfi Zadeh (1965) was the pioneer who first presented the concept of fuzzy set theory through his paper and it is generally agreed that an important point in the assessment of the modern concept of uncertainty was the publication of his seminal paper, even though some thoughts presented in the paper were envisioned 30 years earlier by American philosopher Max Black (1937).

The fuzzy logic is a superset of conventional logic i.e., Boolean logic that handles the concept of partial-truth values between completely true and completely false. That is, propositions are true up to a certain degree and false up to a certain degree. The importance of fuzzy logic derives from the fact that most modes of human reasoning are approximate in nature (Klir and Yuan, 2003).

Applications of Fuzzy Logic:

1. Fuzzy logic can be used for problems associated to uncertainty, imprecise, vagueness, partially true, etc.
2. Fuzzy logic is the best tool for vague human assessments in computing issues.
3. This is also more useful for multiple criteria decision-making problems.
4. The application of fuzzy logic permits the characteristics of the system and their response, a qualitative description of the behaviour of a certain system, without the need for precise mathematical formulation.

From the above, it is concluded that artificial neural network, genetic algorithm, and fuzzy logic can be used for applications in environmental engineering for predictive analysis, optimization problems and uncertainty, and risk and vagueness surrounding the decision-making issues, respectively.

3.4 MULTIPLE CRITERIA DECISION-MAKING (MCDM) MODEL

As we know that fuzzy logic can be used for solving multiple criteria decision-making problems, the procedure of the MCDM model is explained as under for the knowledge of researchers.

3.4.1 MCDM MODEL: AN OVERVIEW OF PROCEDURES

The following steps have been given by Edwards (1977) and Dodgson et al. (2001), which are involved in application of MCDM models:

1. Create the decision context and objectives.
2. Identify the decision-making expert(s) for their domain opinion.
3. Identify the alternatives to be considered for that particular problem.
4. Identify the criteria (parameters) that are related to the decision problem.
5. Assign scores for each of the criteria to measure the performance of the alternatives against each of these criteria and create an evaluation matrix.
6. Standardize the data to generate a priority scores matrix.
7. Determine an importance weightage for each criterion to reflect importance of a particular criterion for the overall decision.
8. Then combine the weights and priority scores by aggregation for the determination of an overall score for each decision alternative. This overall score of alternative forms the basis of a preference ranking.
9. Take a provisional decision on the alternatives, on the basis of preference ranking.

3.4.2 FUZZY MULTIPLE CRITERIA DECISION-MAKING (FMCDM) MODEL

The following developed model (Lad and Christian, 2010) with an illustrative example is explained here as it will be helpful for solving multiple criteria decision-making problems in the environmental engineering field.

According to Hipel et al. (1993), a decision-making problem is said to be difficult and complex; it involves multiple decision makers; multiple criteria, both qualitative and quantitative in nature; uncertainty; vagueness; and risk surrounding the decision making. Using the FMCDM approach, this problem can be solved.

In this study, the FMCDM approach has been developed and explained as follows:

Saaty (1977) has proposed fuzzy numbers related to seven linguistic terms, to define the level of performance on decision criteria. A very few and too many linguistic terms may not provide more help and make the system too complex respectively. The selection of number and type of linguistic terms is a very vital part in decision making issues. Actually, there is no any rule to use four or five or six or nine etc sets of linguistic terms. Only see that selected linguistic terms are simple which to be understood by an expert, system analysts can use easily and real-world application oriented (Chen and Hwang, 1992). In this model, four linguistic terms have been selected that would be convenient for an expert to distinguish subjectively between them.

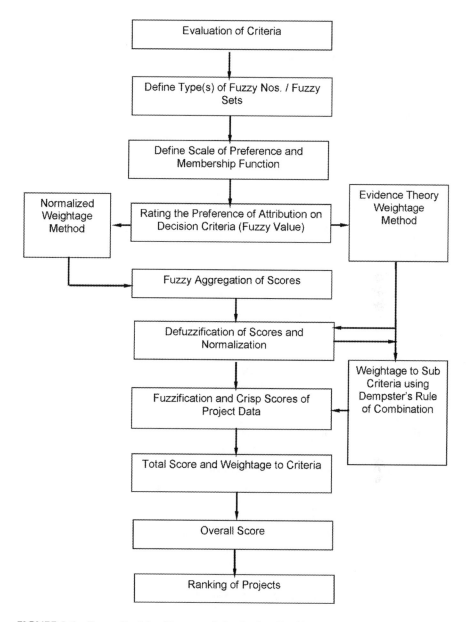

FIGURE 3.1 Fuzzy Decision Framework for Project Ranking.

Figure 3.1 shows the fuzzy decision framework for project ranking (Malczewski, 1999).

Linguistic terms used are: Very Important-VI, Important-I, Average-A, Not Important-NI. and/or Least Important-LI. Table 3.1 shows the fuzzy numbers and linguistic terms. Figure 3.2 shows the fuzzy sets for the selected linguistic terms (Chen and Hwang, 1992).

TABLE 3.1

Fuzzy Numbers and Linguistic Terms

Linguistic Terms	Fuzzy Number
Very Important-VI	(0.72, 0.86, 1.00, 1.00)
Important-I	(0.43, 0.57, 0.72, 0.86)
Average-A	(0.14, 0.29, 0.43, 0.57)
Not Important/Least Important- NI/LI	(0.00, 0.00, 0.14, 0.29)

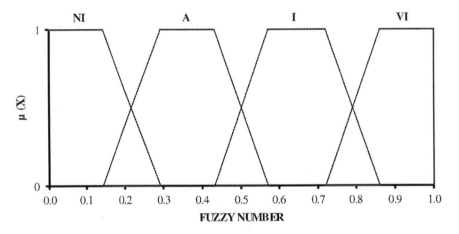

FIGURE 3.2   Fuzzy Numbers for Linguistic Terms.

3.4.2.1   Fuzzy Normalized Weightage Method

The fuzzy normalized weightage method is explained as follows:

> Opinions or perception of experts' (academicians and professionals) in terms of lin-
> guistic terms for sub-criteria (parameters) of that specific work are required to be
> taken. The experts should be selected in such a way that they should be involved in the
> field of particular area for which model is require to be developed.

Then, calculate the average fuzzy numbers from the equation 3.12.

$$A_{ij}^k = (1/p) \cdot \left(a_{i1}^k + a_{i2}^k + \ldots + a_{ip}^k\right) \quad \text{for } i = 1, \ 2, \ \ldots, \ n \text{ and } j = 1, \ 2, \ \ldots, p$$

$$(3.12)$$

Where

a_{ij}^k = fuzzy number (converted from assigned linguistic term by expert) related
     to a sub-criterion (parameter)

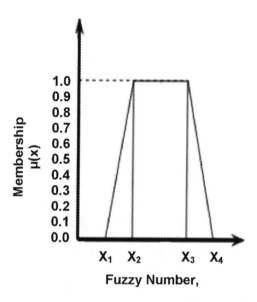

FIGURE 3.3   Trapezoidal Fuzzy Membership Function.

p = number of experts and
n = number of fuzzy numbers

Then, do defuzzification i.e., convert to crisp value. Here, the main aim of crisp value is that it is easy to justify one value of the aggregated fuzzy number. As most of the environmental projects' status is based on range value of parameters, so in this model, trapezoidal fuzzy numbers are used. Figure 3.3 shows the trapezoidal fuzzy membership function (Kaufmann and Gupta, 1991).

For the sub-criteria, the defuzzified value 'e' (Kaufmann and Gupta, 1991) can be calculated as

$$e = (x_1 + x_2 + x_3 + x_4)/4 \qquad (3.13)$$

Then,

Normalized weight for every sub criterion = the crisp score of every sub criterion (C_{mk}) / the summation of crisp score of all sub criteria $\left(\Sigma C_{mk} \right)$

Where

m = the criterion and
k = the sub-criterion.

The next step is standardization of data, i.e., convert the parametric values to the fuzzy numbers based on the specified statutory norms. For example, if bio-chemical

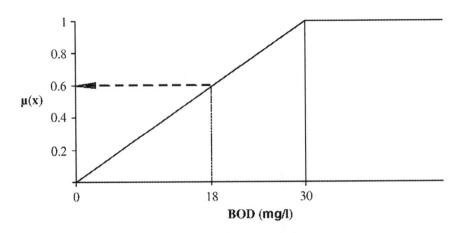

FIGURE 3.4 Pollution Parameter BOD: Fuzzy Set for Not Acceptable.

oxygen demand (BOD) of a given wastewater sample is 18 mg/L and permissible limit of BOD is 30 mg/L, then the fuzzy number (membership function) of that sample is 0.6 (see Figure 3.4) (Lad et al., 2008; Lad and Christian, 2010).

For sub-criterion, a fuzzy decision matrix is as follows:

Where $a_1, a_2, a_3, ..., a_n, b_1, b_2, b_3, ..., b_n$ and $c_1, c_2, c_3, ..., c_n$ are fuzzy values of particular parameter for different projects $P_1, P_2, ..., P_n$, respectively.

The crisp scores of the sub-criterion for each project can be calculated as follows:

Project, $P_1 = (a_1 + a_2 + a_3 + ... + a_n)/n$

Project, $P_2 = (b_1 + b_2 + b_3 + ... + b_n)/n$

Project, $P_n = (c_1 + c_2 + c_3 + ... + c_n)/n$

Now, the total score (TS_{mi}), for each project of different criteria can be calculated by the simple additive weighing method (Hwang and Yoon, 1981) (refer to equation 3.14).

$$TS_{mi} = \sum [X_{mk} \cdot W(C_{mk})] \text{ for } k = 1, \ 2, \ ...n \qquad (3.14)$$

Where,

TS_{mi} = total score of the project

X_{mk} = crisp score of the project data and

$W(C_{mk})$ = weight (importance value)

i = the project

m = the criterion

k = the sub-criterion

Now, an importance weight $[W(C_{mi})]$ can be calculated as

$$W(C_{mi}) = TS_{mi}/\sum TS_{mi} \qquad (3.15)$$

Then, an overall score (OS) for the projects can be calculated using equation (3.16).

$$OS = \sum [TS_{mi} \cdot W(C_{mi})] \text{ for } i = 1, \ 2, \ \ldots n \tag{3.16}$$

The next stage is to do a ranking of projects on the basis of an overall score of projects.

The same concept can also be used by applying weightage obtained from the evidence theory weightage method, as explained as following.

3.4.2.2 Fuzzy Evidence Theory Weightage Method

In the normalized weightage method, the influence of interaction of sub-criteria on parametric weight is not considered. In this method, weightage to every sub-criterion is given as per the relation of sub-criteria with respect to each other; for example, as in the environment the sub-criteria of wastewater discharges and air emissions are expected to react with each other respectively. The evidence theory weightage method is explained as below.

Opinions or perception of experts (academicians and professionals) in terms of linguistic terms for individual sub-criterion and combination of sub-criteria of that specific work are required to be taken. The experts should be selected in such a way that they should be involved in the field of particular area for which model is require to be developed.

The next stage is to calculate the importance weightage factors for these sub-criteria.

Combined evidence can be obtained from two independent sources and expressed by two basic assignments, m_1 and m_2, on some power set.

For different types of projects, experts' opinions, for the relation of all sub-criteria with each other, of that particular project are also required to be obtained separately.

As a basic assignment, for focal elements (separate and combination of sub-criteria), the crisp score i.e., defuzzification of fuzzy numbers, of that a particular linguistic term can be obtained by using equation (3.13) and then basic assignments m_1 and m_2 for each focal element of different criteria can be calculated as the crisp score of each focal element (C_{mk}) divided by the summation of all focal elements ($\sum C_{mk}$).

The next stage is to appropriately combine two basic assignments (m_1 and m_2) on some power set to obtain a joint basic assignment, $m_{1,2}$.

The appropriate way of combining evidence is Dempster's rule of combination and the same is expressed by equation 3.17.

$$m_{1,2}(A) = \frac{\sum_{B \cap C = A} m_1(B) \cdot m_2(C)}{1 - K} \tag{3.17}$$

for all $A \neq \phi$ and $m_{1,2}(\phi) = 0$, where

$$K = \sum_{B \cap C = \phi} m_1(B) \cdot m_2(C) \tag{3.18}$$

As per this rule, the degree of evidence, $m_1(B)$, from the first source that focuses on set $B \in P(X)$ and the degree of evidence $m_2(C)$ from the second source that focuses on set $C \in P(X)$ can be combined by taking the product $m_1(B) \cdot m_2(C)$, which focuses on the intersection $B \cap C$. This is exactly the same way in which the joint probability distribution can be calculated from two independent marginal distributions; consequently, it is justified on the same grounds. However, since some intersections of focal elements from the first $[m_1(B)]$ and second $[m_2(C)]$ source may result in the same set A, it is a must to add the corresponding products to obtain $m_{1,2}(A)$. Moreover, some of the intersections may be empty. Since it is required that $m_{1,2}(\phi) = 0$, the value K is not included in the definition of the joint basic assignment $m_{1,2}$. This means that the sum of products $m_1(B) \cdot m_2(C)$ for all focal elements B of m_1 and all focal elements C of m_2 such that $B \cap C \neq \phi$ is equal to $(1 - K)$. To obtain a normalized basic assignment $m_{1,2}$ that is $\sum_{A \in P(X)} m(A) = 1$ it is required to divide each of these products by factor $(1 - K)$. The $m_{1,2}$ obtained from the above equation for each sub-criterion of different criteria is the normalized weight (Klir and Yuan, 2003).

3.4.2.3   Illustrative Example

The application of the above fuzzy multiple criteria decision-making model is explained through the following illustrative example of evaluation of environmental pollution potential index (EPPI) for industries (Lad and Christian, 2010).

a. Evaluation of Environmental Pollution Potential Index (EPPI) for Industries

For the determination of EPPI, the fuzzy MCDM approach is used. The normalized weight for each sub-criterion of water and air pollution is calculated by two methods, as shown below:

> The parametric values of wastewater discharges and stack emissions of three Chemical, three Textile, three Engineering (Automobile) and three Dairy industries and the permissible limits for industrial effluents and stack emissions Prescribed by Maharashtra Pollution Control Board are considered for this example.

Fuzzy Normalized Weightage Method

Opinions or perception of experts (academicians and professionals) in terms of linguistic terms have been taken for sub-criteria (parameters), who are involved in the field of environmental engineering.

Opinions or perception of experts (linguistic terms) for 26 sub-criteria of water pollution (suspended solids, colour, TDS, temperature, BOD, COD, oil and grease, chlorides, sulphates, ammonia nitrogen, phosphates, nickel, phenolic compounds, sulphides, fluorides, zinc, arsenic, cadmium, chromium, copper, cyanides, selenium, lead, mercury, iron, and pH for wastewater) and for five sub-criteria of air pollution (SPM, SO_2, NOx, Cl_2, and HCl) have been taken from academicians and professionals. For example, for suspended solids expert opinion in terms of linguistic terms have been given by five academicians like I,I,A,A,A, respectively, and by five professionals like A,A,I,VI,I, respectively.

Then opinions or perception of experts in terms of linguistic terms of every sub-criterion (parameter) of water and air pollution are changed to fuzzy numbers using Table 3.1 and Figure 3.4 and then using equations as shown below, the average fuzzy numbers and crisp score, respectively, for each parameter of water and air pollution of chemical, textile, engineering (automobile), and dairy industries are calculated using equations 3.12 and 3.13.

As a sample calculation and explanation of the whole procedure, the chemical industry 1 data and academician's opinion are considered.

The fuzzy decision matrix (F_{C_2}), the average fuzzy score matrix (AF_{C_2}), and crisp score for sub-criteria of air pollution of chemical industry 1 are shown as below.

Fuzzy decision matrix (academicians):

$$F_{C_2} = \begin{bmatrix} (0.43,\ 0.57,\ 0.72,\ 0.86)(0.43,\ 0.57,\ 0.72,\ 0.86)(0.72,\ 0.86,\ 1.00,\ 1.00) \\ (0.72,\ 0.86,\ 1.00,\ 1.00)(0.72,\ 0.86,\ 1.00,\ 1.00) \\ (0.72,\ 0.86,\ 1.00,\ 1.00)(0.72,\ 0.86,\ 1.00,\ 1.00)(0.72,\ 0.86,\ 1.00,\ 1.00) \\ (0.72,\ 0.86,\ 1.00,\ 1.00)(0.72,\ 0.86,\ 1.00,\ 1.00) \\ (0.43,\ 0.57,\ 0.72,\ 0.86)(0.72,\ 0.86,\ 1.00,\ 1.00)(0.72,\ 0.86,\ 1.00,\ 1.00) \\ (0.14,\ 0.29,\ 0.43,\ 0.57)(0.72,\ 0.86,\ 1.00,\ 1.00) \\ (0.43,\ 0.57,\ 0.72,\ 0.86)(0.43,\ 0.57,\ 0.72,\ 0.86)(0.43,\ 0.57,\ 0.72,\ 0.86) \\ (0.43,\ 0.57,\ 0.72,\ 0.86)(0.43,\ 0.57,\ 0.72,\ 0.86) \\ (0.14,\ 0.29,\ 0.43,\ 0.57)(0.14,\ 0.29,\ 0.43,\ 0.57)(0.14,\ 0.29,\ 0.43,\ 0.57) \\ (0.14,\ 0.29,\ 0.43,\ 0.57)(0.43,\ 0.57,\ 0.72,\ 0.86) \end{bmatrix}$$

Average fuzzy score matrix (academicians):

$$AF_{C_2} = \begin{bmatrix} 0.604,\ 0.744,\ 0.888,\ 0.944 \\ 0.720,\ 0.860,\ 1.000,\ 1.000 \\ 0.546,\ 0.688,\ 0.830,\ 0.886 \\ 0.430,\ 0.570,\ 0.720,\ 0.860 \\ 0.198,\ 0.346,\ 0.488,\ 0.628 \end{bmatrix} \begin{matrix} SPM \\ SO_2 \\ NO_x \\ Cl_2 \\ HCl \end{matrix}$$

Crisp scores:

Criterion, C_{21}(SPM) $= (0.604 + 0.744 + 0.888 + 0.944)/4 = 0.795$
Criterion, C_{22}(SO$_2$) $= (0.720 + 0.860 + 1.000 + 1.000)/4 = 0.895$
Criterion, C_{23}(NO$_x$) $= (0.546 + 0.688 + 0.830 + 0.886)/4 = 0.738$
Criterion, C_{24}(Cl$_2$) $= (0.430 + 0.570 + 0.720 + 0.860)/4 = 0.645$
Criterion, C_{25}(HCl) $= (0.198 + 0.346 + 0.488 + 0.628)/4 = 0.415$

Average fuzzy numbers (AFNs) and crisp score, respectively, for each sub-criterion of water and air pollution of chemical industry are as shown in Table 3.2.

Then, the normalized weight for each sub-criterion of water and air pollution was obtained by dividing the score of each sub-criterion (C_{mk}) by the sum total of all sub-criteria (ΣC_{mk}) for different types of industries. The weights for each sub-criteria depended upon the characteristics of wastewater discharges and air emissions from industry. The normalized weight for each sub-criterion of water and air pollution of chemical industry is as shown in Table 3.3.

The parametric data of wastewater discharges and stack emissions are converted to the fuzzy numbers (membership functions) based on the specified statutory norms (see Figure 3.4).

The next step is to determine the total score. To obtain the total score, the fuzzy crisp scores of data and the normalized weight of sub-criteria are operated by a matrix, as shown below.

TABLE 3.2
Average Fuzzy Number (Academicians)

Type of Industry	Sub-Criteria	AFN1	AFN2	AFN3	AFN4	AVG(C_k)
Chemical	**Water Pollution**					
	SS, mg/L	0.256	0.402	0.546	0.686	0.473
	TDS, mg/L	0.488	0.628	0.776	0.888	0.695
	BOD, mg/L	0.720	0.860	1.000	1.000	0.895
	COD, mg/L	0.720	0.860	1.000	1.000	0.895
	Oil and grease, mg/L	0.256	0.402	0.546	0.686	0.473
	Chlorides, mg/L	0.256	0.402	0.546	0.686	0.473
	Sulphates, mg/L	0.256	0.402	0.546	0.686	0.473
	Ammonia nitrogen, mg/L	0.028	0.058	0.198	0.346	0.158
	Nickel, mg/L	0.314	0.458	0.604	0.744	0.530
	Phenolic compounds, mg/L	0.084	0.174	0.314	0.458	0.258
	Sulphides, mg/L	0.256	0.402	0.546	0.686	0.473
	Copper, mg/L	0.198	0.346	0.488	0.628	0.415
	pH	0.198	0.346	0.488	0.628	0.415
	ΣC_k					**6.626**
	Air Pollution					
	SPM, mg/Nm^3	0.604	0.744	0.888	0.944	0.795
	SO$_2$, ppm	0.720	0.860	1.000	1.000	0.895
	NOx, ppm	0.546	0.688	0.830	0.886	0.738
	Cl$_2$, ppm	0.430	0.570	0.720	0.860	0.645
	HCl, mg/Nm^3	0.198	0.346	0.488	0.628	0.415
	ΣC_k					**3.488**

TABLE 3.3
Normalized Weight (Academicians)

Type of Industry	Sub-Criteria	Weight
Chemical	**Water Pollution**	
	SS, mg/L	0.071
	TDS, mg/L	0.105
	BOD, mg/L	0.135
	COD, mg/L	0.135
	Oil and grease, mg/L	0.071
	Chlorides, mg/L	0.071
	Sulphates, mg/L	0.071
	Ammonia nitrogen, mg/L	0.024
	Nickel, mg/L	0.080
	Phenolic compounds, mg/L	0.039
	Sulphides, mg/L	0.071
	Copper, mg/L	0.063
	pH	0.063
	Air Pollution	
	SPM, mg/Nm^3	0.228
	SO$_2$, ppm	0.257
	NOx, ppm	0.211
	Cl$_2$, ppm	0.185
	HCl, mg/Nm^3	0.119

i. Total Score Matrix for Chemical Industries (Academicians)
a. Water Pollution

$$TS_{Water} = \begin{bmatrix} 1.000 & 1.000 & 1.000 \\ 1.000 & 1.000 & 1.000 \\ 1.000 & 1.000 & 1.000 \\ 1.000 & 1.000 & 1.000 \\ 1.000 & 1.000 & 1.000 \\ 1.000 & 1.000 & 1.000 \\ 0.995 & 1.000 & 1.000 \\ 0.000 & 0.000 & 0.000 \\ 0.000 & 0.000 & 0.000 \\ 0.000 & 0.000 & 0.000 \\ 0.000 & 0.000 & 0.000 \\ 0.000 & 0.000 & 0.000 \\ 0.152 & 0.099 & 0.109 \end{bmatrix} \begin{bmatrix} 0.071 \\ 0.105 \\ 0.135 \\ 0.135 \\ 0.071 \\ 0.071 \\ 0.071 \\ 0.024 \\ 0.080 \\ 0.039 \\ 0.071 \\ 0.063 \\ 0.063 \end{bmatrix}$$

I_1 I_2 I_3 W_{C_k}

SS, TDS, BOD, COD, Oil & Grease, Chlorides, Sulphates, Ammoniacal Nitrogen, Nickel, Phenolic Compounds, Sulphides, Copper, pH

Using a simple additive weighing method (Hwang and Yoon, 1981), the total score (TS) for each industry of water and air pollution criteria were calculated separately using the equation shown below, with usual notations.

$$TS_{mi} = \sum [X_{mk} \cdot W(C_{mk})] \text{ for } k = 1, \ 2, \ \dots n$$

As a sample calculation, the total score for sub-criteria of air pollution due to chemical industry 1 (academicians) is as shown below.

$$TS_{Chemical\ 1,Air} = (1.000 \times 0.228 + 1.000 \times 0.257 + 1.000 \times 0.211$$
$$+ 0.995 \times 0.185 + 0.719 \times 0.119)$$
$$= 0.966.$$

The total score for sub-criteria of water and air pollution of all chemical, textile, engineering (automobile), and dairy industries are as shown in Table 3.4.

To determine the pollution potential weightage for both the criteria of water and air pollution (such that their summation was equal to 1), the equation given below was used and the calculated weights are as shown in Table 3.4.

$$W(C_{mi}) = TS_{mi}/\sum TS_{mi}$$

The next step is to determine an overall score. To obtain an overall score, the total score and the weightage of criteria are operated by a matrix, as shown below.

TABLE 3.4
Total Score (Academicians)

Type of Industry	Ind. No.	Total Score		Sum (C1+C2)	Weight	
		Water Pollution (C1)	Air Pollution (C2)		Water Pollution	Air Pollution
Chemical	1	0.670	0.966	1.636	0.410	0.590
	2	0.667	1.000	1.667	0.400	0.600
	3	0.667	0.881	1.548	0.431	0.569
Textile	4	0.674	0.675	1.349	0.500	0.500
	5	0.662	0.652	1.314	0.504	0.496
	6	0.680	0.691	1.371	0.496	0.504
Engineering (Automobile)	7	0.248	0.360	0.608	0.408	0.592
	8	0.291	0.368	0.659	0.441	0.559
	9	0.319	0.425	0.744	0.429	0.571
Dairy	10	0.518	0.457	0.975	0.531	0.469
	11	0.513	0.480	0.993	0.517	0.483
	12	0.536	0.515	1.051	0.510	0.490

Overall Score Matrix for Chemical Industries (Academicians):

Using the simple additive weighing method (Hwang and Yoon, 1981), the overall score (OS) is

$$OS_{I_1} = \begin{bmatrix} 0.670 \\ 0.966 \end{bmatrix} \begin{bmatrix} 0.410 \\ 0.590 \end{bmatrix}; \quad OS_{I_2} = \begin{bmatrix} 0.667 \\ 1.000 \end{bmatrix} \begin{bmatrix} 0.400 \\ 0.600 \end{bmatrix}; \quad OS_{I_3} = \begin{bmatrix} 0.667 \\ 0.881 \end{bmatrix} \begin{bmatrix} 0.431 \\ 0.569 \end{bmatrix} \begin{matrix} \text{Water} \\ \text{Air} \end{matrix}$$

for the industries are calculated using equation as shown below, with usual notations.

$$OS = \sum [TS_{mi} \cdot W(C_{mi})] \text{ for } i = 1, 2, \dots n$$

As a sample calculation, an overall score for complex water and air pollution criteria of chemical industry 1 (academicians) is as shown below.

$$OS_{Chemical1} = (0.670 \times 0.410 + 0.966 \times 0.590)$$
$$= 0.844.$$

Table 3.5 shows the overall score and ranking of all chemical, textile, engineering (automobile), and dairy industries.

TABLE 3.5
Overall Score and Ranking (Academicians and Professionals)

Type of Industry	Ind. No.	Overall Score		Rank
		Academicians	Professionals	
Chemical	1	0.844	0.840	2
	2	0.867	0.869	1
	3	0.789	0.769	3
Textile	4	0.674	0.658	5
	5	0.657	0.645	6
	6	0.686	0.671	4
Engineering (Automobile)	7	0.314	0.301	12
	8	0.334	0.321	11
	9	0.380	0.357	10
Dairy	10	0.489	0.468	9
	11	0.497	0.476	8
	12	0.526	0.500	7

From the normalized weightage method, the following points are observed:

i. The final ranking of industries did not change with the linguistic opinion of the experts (academicians and professionals). However, the EPPI for an industry marginally changed. This was mainly due to the change in the normalized weightage factors derived on the basis of the linguistic term assignment by the experts.

ii. The water pollution potential is more than the air pollution potential due to the dairy industries. The air pollution potential is more than the water pollution potential due to the chemical and engineering (automobile) industries. The water and air pollution potential due to the textile industries is almost the same in bothdomains of the environment.

iii. Industry 2 ranked first out of 12 industries and it showed higher pollution potential, while industry 7 ranked number 12 with the lowest pollution potential compared to other industries.

iv. The complex EPPI of chemical, textile, engineering (automobile), and dairy industries is in the range of 0.769 to 0.869, 0.645 to 0.686, 0.301 to 0.380, and 0.468 to 0.526, respectively (see Table 3.5).

v. Table 3.5 reflects that the chemical industry has the highest pollution potential followed by textile and dairy industries. The engineering (automobile) industry is the least polluting type of industry.

ii. Fuzzy Evidence Theory Weightage Method

Opinions or perceptions of experts (academicians and professionals) in terms of linguistic terms have been taken for individual sub-criterion and combination of sub-criteria of water and air pollution for different types of industries.

Opinions or perception of experts, for the relation of all sub-criteria with each other, of that particular industry, of water and air pollution, are taken separately.

The next stage is mass assignment for focal elements is calculated. As a sample, Table 3.6 shows a mass assignment for focal elements of air pollution due to the textile industry.

Similarly, a mass assignment for each focal element of water pollution of the textile industry and each focal element of water and air pollution of the chemical, engineering (automobile), and dairy industries are calculated with consideration of a combination of all opinions of experts with each other.

Applying the following equation of Dempster's rule to m_1 and m_2, combined evidence (joint basic assignment) $m_{1,2}$, is obtained.

$$m_{1,2}(A) = \frac{\sum_{B \cap C = A} m_1(B) \cdot m_2(C)}{1 - K}$$

To determine the values of $m_{1,2}$, firstly, the normalization factor $(1 - K)$ is calculated. Applying the equation as shown below, with usual notations, K is obtained as follows:

$$K = \sum_{B \cap C = \phi} m_1(B) \cdot m_2(C)$$

TABLE 3.6

Mass Assignment for Focal Elements of Air Pollution (Academicians)

Focal Elements	Expert1			Expert2		
	Ling. Term	Crisp Score	Mass 1 (m_1)	Ling. Term	Crisp Score	Mass 2 (m_2)
Air Pollution						
SPM	I	0.645	0.2952	I	0.645	0.2648
SO_2	V I	0.895	0.4096	V I	0.895	0.3676
NOx	I	0.645	0.2952	V I	0.895	0.3676
SPM $\cup$ SO_2	N S	0	0	N S	0	0
SPM $\cup$ NOx	N S	0	0	N S	0	0
SO_2 $\cup$ NOx	N S	0	0	N S	0	0
SPM $\cup$ $SO_2 \cup$ NOx	N S	0	0	N S	0	0
TOTAL		$\Sigma\,C_{mk}$ = **2.185**	**1**		$\Sigma\,C_{mk}$ = **2.435**	**1**

$$K = m_1(SPM) . m_2(SO_2) + m_1(SPM) . m_2(NOx) + m_1(SPM) . m_2(SO_2 \cup NOx)$$
$$+ m_1(SO_2) . m_2(SPM) + m_1(SO_2) . m_2(NOx) + m_1(SO_2) . m_2(SPM \cup NOx)$$
$$+ m_1(NOx) . m_2(SPM) + m_1(NOx) . m_2(SO_2) + m_1(NOx) . m_2(SPM \cup SO_2)$$
$$+ m_1(SPM \cup SO_2) . m_2(NOx) + m_1(SPM \cup NOx) . m_2(SO_2)$$
$$+ m_1(SO_2 \cup NOx) . m_2(SPM)$$
$$= 0.663$$

The normalization factor is then $(1 - K) = 0.337$.

Values of combined evidence $m_{1,2}$ are calculated by equation (3.6). For example,

$$m_{1,2}(SPM) = [m_1(SPM) . m_2(SPM) + m_1(SPM) . m_2(SPM \cup SO_2) + m_1(SPM)$$
$$\times m_2(SPM \cup NOx) + m_1(SPM) . m_2(SPM \cup SO_2 \cup NOx)$$
$$+ m_1(SPM \cup SO_2) . m_2(SPM) + m_1(SPM \cup SO_2) . m_2(SPM \cup NOx)$$
$$+ m_1(SPM \cup NOx) . m_2(SPM) + m_1(SPM \cup NOx) . m_2(SPM \cup SO_2)$$
$$+ m_1(SPM \cup SO_2 \cup NOx) . m_2(SPM)]/0.337$$
$$= 0.3218$$

Similarly, values of combined evidence ($m_{1,2}$) for the remaining focal elements SO_2 and NOx are calculated and the same are as shown in Table 3.7.

Then, combined evidence ($m_{1,2}$), by considering combination of mass assignments from two independent sources (Academicians: Expert1 and Expert2), for each sub-criterion of water pollution of the textile industry and for each sub-criterion of water and air pollution of chemical, engineering (automobile), and dairy industries is calculated.

Table 3.8 shows the combined evidence calculated by considering a combination of mass assignments from two independent sources (Academicians: Expert1 and

TABLE 3.7

Combination of Degrees of Evidence from Two Independent Sources (Academicians: Expert1 and Expert2)

Focal Elements	Expert1 m_1	Expert2 m_2	Combined Evidence $m_{1,2}$
SPM	0.2952	0.2648	0.3218
SO_2	0.4096	0.3676	0.4464
NOx	0.2952	0.3676	0.2318

TABLE 3.8

Combination of Degrees of Evidence from Two Independent Sources (Academicians: Expert1 and Expert2)

Sr. No.	Type of Industry	Focal Elements	Expert1 m_1	Expert2 m_2	Combined Evidence $m_{1,2}$
1	Chemical	**Water Pollution**			
		SS, mg/L	0.0415	0.0383	0.041
		TDS, mg/L	0.0415	0.0383	0.088
		BOD, mg/L	0.0576	0.0531	0.067
		COD, mg/L	0.0576	0.0531	0.156
		Oil and grease, mg/L	0.0415	0.0213	0.010
		Chlorides, mg/L	0.0230	0.0213	0.025
		Sulphates, mg/L	0.0230	0.0383	0.022
		Ammonia nitrogen, mg/L	0.0069	0.0064	0.004
		Nickel, mg/L	0.0415	0.0383	0.016
		Phenolic compounds, mg/L	0.0230	0.0213	0.011
		Sulphides, mg/L	0.0230	0.0213	0.017
		Copper, mg/L	0.0230	0.0213	0.036
		pH	0.0230	0.0383	0.005
		Air Pollution			
		SPM, mg/Nm^3	0.2023	0.1876	0.178
		SO_2, ppm	0.2807	0.2603	0.342
		NOx, ppm	0.2023	0.2603	0.247
		Cl_2, ppm	0.2023	0.1876	0.178
		HCl, mg/Nm^3	0.1123	0.1041	0.055
		SS, mg/L	0.0703	0.0564	0.140
		Colour, Co-pt	0.0390	0.0313	0.050

Expert2), for each sub-criterion of water and air pollution of the chemical, textile, engineering (automobile), and dairy industries.

The combined evidence, by considering combination of mass assignments from other experts, for each sub-criterion of water and air pollution of the chemical, textile, engineering (automobile), and dairy industries is calculated. Table 3.9 shows the combined evidence from all experts for each sub-criterion of water and air pollution of the chemical industry.

Table 3.10 shows the combined evidence from all experts for each sub-criterion of water and air pollution of the textile industry.

Table 3.11 shows the combined evidence from all experts for each sub-criterion of water and air pollution of the engineering (automobile) industry.

Table 3.12 shows the combined evidence from all experts for each sub-criterion of water and air pollution of the dairy industry.

The next step is to determine the total score. To obtain the total score, the fuzzy crisp scores of data and the normalized weight of sub-criteria are operated by a matrix, as shown below.

i. Matrix for Chemical Industries (Academicians)
a. Water Pollution

$$
TS_{Water} = \begin{array}{cccc}
I_1 & I_2 & I_3 & W_{C_k}
\end{array}
\begin{bmatrix}
1.000 & 1.000 & 1.000 \\
1.000 & 1.000 & 1.000 \\
1.000 & 1.000 & 1.000 \\
1.000 & 1.000 & 1.000 \\
1.000 & 1.000 & 1.000 \\
1.000 & 1.000 & 1.000 \\
0.995 & 1.000 & 1.000 \\
0.000 & 0.000 & 0.000 \\
0.000 & 0.000 & 0.000 \\
0.000 & 0.000 & 0.000 \\
0.000 & 0.000 & 0.000 \\
0.000 & 0.000 & 0.000 \\
0.152 & 0.099 & 0.109
\end{bmatrix}
\begin{bmatrix}
0.041 \\
0.088 \\
0.067 \\
0.156 \\
0.010 \\
0.025 \\
0.022 \\
0.004 \\
0.016 \\
0.011 \\
0.017 \\
0.036 \\
0.005
\end{bmatrix}
\begin{array}{l}
\text{SS} \\
\text{TDS} \\
\text{BOD} \\
\text{COD} \\
\text{Oil \& Grease} \\
\text{Chlorides} \\
\text{Sulphates} \\
\text{Ammoniacal Nitrogen} \\
\text{Nickel} \\
\text{Phenolic Compounds} \\
\text{Sulphides} \\
\text{Copper} \\
\text{pH}
\end{array}
$$

Table 3.13 shows the overall score and ranking of all the chemical, textile, engineering (automobile), and dairy industries.

Similarly, the ranking of all chemical, textile, engineering (automobile), and dairy industries is calculated on the basis of combination of opinions of other experts. Table 3.14 shows the overall score and ranking of all industries on the basis of combination of all opinions of experts with each other.

From the evidence theory weightage method, it is observed that:

    i. The ranking of industries by normalized and evidence theory weightage method did not change with the opinion of the experts.

    ii. In the evidence theory weightage method, importance weightage is considered for separate and combination of sub-criteria of water and air

TABLE 3.9

Combined Evidence from all Experts (Academicians)

Type of Industry	Sub-Criteria	Combined Evidence									
		$m_{1,2}$	$m_{1,3}$	$m_{1,4}$	$m_{1,5}$	$m_{2,3}$	$m_{2,4}$	$m_{2,5}$	$m_{3,4}$	$m_{3,5}$	$m_{4,5}$
Chemical	**Water Pollution**										
	SS	0.041	0.036	0.033	0.037	0.034	0.032	0.036	0.027	0.031	0.029
	TDS	0.088	0.149	0.100	0.146	0.088	0.058	0.086	0.099	0.144	0.097
	BOD	0.067	0.079	0.074	0.078	0.055	0.051	0.054	0.061	0.064	0.060
	COD	0.156	0.110	0.146	0.110	0.169	0.218	0.167	0.158	0.119	0.157
	O & G	0.010	0.014	0.010	0.011	0.010	0.006	0.007	0.009	0.010	0.007
	Chlorides	0.025	0.025	0.026	0.027	0.023	0.021	0.024	0.024	0.025	0.025
	Sulphates	0.022	0.019	0.016	0.015	0.027	0.023	0.023	0.020	0.020	0.017
	Amm. Nitrogen	0.004	0.005	0.007	0.005	0.003	0.005	0.003	0.006	0.004	0.006
	Nickel	0.016	0.013	0.012	0.010	0.015	0.011	0.011	0.011	0.009	0.008
	Phe. Compounds	0.011	0.006	0.008	0.010	0.008	0.008	0.012	0.006	0.007	0.009
	Sulphides	0.017	0.023	0.016	0.024	0.025	0.016	0.026	0.023	0.035	0.025
	Copper	0.036	0.033	0.034	0.030	0.049	0.047	0.045	0.047	0.041	0.042
	pH	0.005	0.003	0.003	0.003	0.005	0.004	0.005	0.003	0.003	0.003
	Air Pollution										
	SPM	0.178	0.231	0.268	0.222	0.212	0.257	0.204	0.325	0.263	0.312
	SO_2	0.342	0.320	0.372	0.308	0.294	0.357	0.284	0.325	0.263	0.312
	NOx	0.247	0.231	0.107	0.222	0.294	0.143	0.284	0.130	0.263	0.125
	Cl_2	0.178	0.166	0.193	0.160	0.153	0.185	0.147	0.169	0.137	0.162
	HCl	0.055	0.051	0.059	0.089	0.047	0.057	0.082	0.052	0.076	0.090

TABLE 3.10

Combined Evidence from all Experts (Academicians)

Type of Industry	Sub-Criteria	Combined Evidence									
		$m_{1,2}$	$m_{1,3}$	$m_{1,4}$	$m_{1,5}$	$m_{2,3}$	$m_{2,4}$	$m_{2,5}$	$m_{3,4}$	$m_{3,5}$	$m_{4,5}$
Textile	**Water Pollution**										
	SS	0.140	0.137	0.111	0.134	0.118	0.093	0.116	0.091	0.111	0.089
	Colour	0.050	0.037	0.046	0.034	0.046	0.058	0.042	0.042	0.032	0.038
	BOD	0.092	0.126	0.102	0.127	0.084	0.067	0.084	0.093	0.117	0.094
	COD	0.195	0.183	0.204	0.188	0.211	0.223	0.211	0.221	0.204	0.222
	O & G	0.029	0.052	0.032	0.042	0.026	0.015	0.021	0.029	0.039	0.022
	Phe. Compounds	0.024	0.013	0.020	0.022	0.017	0.026	0.028	0.013	0.014	0.023
	Sulphides	0.039	0.048	0.041	0.052	0.053	0.045	0.058	0.055	0.070	0.060
	Chromium	0.033	0.031	0.034	0.034	0.036	0.039	0.039	0.037	0.036	0.040
	pH	0.011	0.010	0.008	0.011	0.010	0.008	0.011	0.008	0.010	0.008
	Air Pollution										
	SPM	0.232	0.295	0.359	0.295	0.265	0.340	0.265	0.417	0.333	0.417
	SO_2	0.447	0.409	0.498	0.409	0.368	0.472	0.368	0.417	0.333	0.417
	NOx	0.322	0.295	0.144	0.295	0.368	0.189	0.368	0.167	0.333	0.167

TABLE 3.11

Combined Evidence from All Experts (Academicians)

Type of Industry	Sub-Criteria	Combined Evidence									
		$m_{1,2}$	$m_{1,3}$	$m_{1,4}$	$m_{1,5}$	$m_{2,3}$	$m_{2,4}$	$m_{2,5}$	$m_{3,4}$	$m_{3,5}$	$m_{4,5}$
Engg. (Automobile)	**Water Pollution**										
	SS	0.043	0.036	0.035	0.038	0.034	0.032	0.036	0.026	0.030	0.028
	TDS	0.101	0.167	0.115	0.166	0.103	0.071	0.102	0.118	0.169	0.116
	BOD	0.071	0.078	0.074	0.077	0.058	0.055	0.057	0.061	0.063	0.060
	COD	0.180	0.137	0.166	0.138	0.201	0.236	0.201	0.186	0.155	0.187
	O & G	0.012	0.017	0.013	0.013	0.011	0.008	0.008	0.012	0.012	0.009
	Chlorides	0.037	0.045	0.040	0.048	0.035	0.030	0.037	0.038	0.044	0.040
	Sulphates	0.034	0.028	0.026	0.024	0.042	0.037	0.037	0.033	0.031	0.028
	Nickel	0.019	0.015	0.016	0.012	0.018	0.015	0.014	0.015	0.011	0.011
	Zinc	0.030	0.027	0.028	0.023	0.040	0.039	0.035	0.039	0.032	0.033
	Chromium	0.019	0.018	0.019	0.019	0.021	0.019	0.022	0.020	0.022	0.021
	pH	0.004	0.003	0.003	0.003	0.004	0.005	0.004	0.003	0.003	0.003
	Air Pollution										
	SPM	0.232	0.295	0.359	0.295	0.265	0.340	0.265	0.417	0.333	0.417
	SO_2	0.447	0.409	0.498	0.409	0.368	0.472	0.368	0.417	0.333	0.417
	NOx	0.322	0.295	0.144	0.295	0.368	0.189	0.368	0.167	0.333	0.167

TABLE 3.12
Combined Evidence from All Experts (Academicians)

Type of Industry	Sub-Criteria	Combined Evidence									
		$m_{1,2}$	$m_{1,3}$	$m_{1,4}$	$m_{1,5}$	$m_{2,3}$	$m_{2,4}$	$m_{2,5}$	$m_{3,4}$	$m_{3,5}$	$m_{4,5}$
Dairy	**Water Pollution**										
	SS	0.089	0.067	0.067	0.063	0.074	0.074	0.070	0.055	0.052	0.052
	Colour	0.024	0.017	0.020	0.019	0.027	0.032	0.030	0.023	0.021	0.024
	TDS	0.106	0.139	0.119	0.146	0.118	0.100	0.122	0.133	0.161	0.138
	BOD	0.095	0.107	0.107	0.105	0.081	0.081	0.079	0.090	0.088	0.088
	COD	0.143	0.123	0.134	0.122	0.150	0.161	0.148	0.141	0.129	0.139
	O & G	0.019	0.024	0.018	0.019	0.018	0.013	0.014	0.017	0.018	0.013
	Chlorides	0.022	0.033	0.025	0.034	0.022	0.017	0.023	0.024	0.032	0.025
	Sulphates	0.019	0.017	0.012	0.012	0.026	0.020	0.021	0.017	0.018	0.013
	Amm. Nitrogen	0.006	0.007	0.010	0.007	0.005	0.008	0.006	0.010	0.007	0.010
	Sulphides	0.022	0.022	0.021	0.023	0.031	0.027	0.033	0.030	0.033	0.031
	pH	0.008	0.005	0.005	0.005	0.007	0.007	0.008	0.005	0.005	0.005
	Air Pollution										
	SPM	0.232	0.295	0.359	0.295	0.265	0.340	0.265	0.417	0.333	0.417
	SO_2	0.447	0.409	0.498	0.409	0.368	0.472	0.368	0.417	0.333	0.417
	NOx	0.322	0.295	0.144	0.295	0.368	0.189	0.368	0.167	0.333	0.167

TABLE 3.13

Overall Score and Ranking (Academicians and Professionals: Expert1 and Expert2)

Type of Industry	Ind. No.	Overall Score		Rank
		Academicians	Professionals	
Chemical	1	0.814	0.811	2
	2	0.828	0.829	1
	3	0.783	0.769	3
Textile	4	0.578	0.564	5
	5	0.567	0.544	6
	6	0.589	0.570	4
Engineering (Automobile)	7	0.286	0.296	12
	8	0.292	0.298	11
	9	0.323	0.317	10
Dairy	10	0.391	0.380	9
	11	0.399	0.392	8
	12	0.425	0.417	7

pollution, respectively. In the case of water pollution sub-criteria, weightage is distributed among individual sub-criterion and combination of sub-criteria and so weightage to individual sub-criterion is reduced comparatively than normalized weightage method. In the case of air pollution, according to the experts' perception, there is no relation of sub-criteria to each other, so the importance weightage of individual sub-criteria did not differ significantly from the normalized weightage method. Therefore, in evidence theory weightage method, weightage to air pollution criteria is more than water pollution criteria.

iii. the As the highest pollution potential, Industry 2 ranked first out of 12 industries and as lowest pollution potential industry 7 ranked number 12.

iv. Table 3.14 and Table 3.15 reflect that the chemical industry is the highest pollution potential, followed by the textile and dairy industries. The engineering (automobile) industry is the least polluting type of industry.

From the normalized and evidence theory weightage methods, the following points are observed:

i. From the fuzzy multiple criteria decision-making (FMCDM) method, it can be possible to rank the industries based on EPPI.

ii. From the fuzzy multiple criteria decision-making method, water and air pollution potential can be calculated separately.

iii. Ranking by both methods is the same.

TABLE 3.14

Overall Score and Ranking (Academicians)

Type of Industry	Ind. No.	Overall Score and combination of Mass										Rank
		1,2	1,3	1,4	1,5	2,3	2,4	2,5	3,4	3,5	4,5	
Chemical	1	0.814	0.816	0.813	0.808	0.817	0.814	0.809	0.816	0.811	0.807	2
	2	0.828	0.829	0.828	0.829	0.828	0.828	0.829	0.829	0.830	0.829	1
	3	0.783	0.787	0.779	0.757	0.789	0.781	0.761	0.785	0.767	0.755	3
Textile	4	0.578	0.597	0.632	0.598	0.570	0.614	0.570	0.635	0.592	0.634	5
	5	0.567	0.585	0.606	0.586	0.561	0.590	0.560	0.609	0.580	0.609	6
	6	0.589	0.610	0.644	0.610	0.580	0.625	0.580	0.647	0.603	0.646	4
Engg. (Automobile)	7	0.286	0.301	0.366	0.301	0.269	0.348	0.269	0.360	0.287	0.360	12
	8	0.292	0.310	0.372	0.310	0.277	0.352	0.277	0.367	0.298	0.367	11
	9	0.323	0.351	0.422	0.351	0.314	0.400	0.314	0.428	0.346	0.428	10
Dairy	10	0.391	0.406	0.457	0.404	0.374	0.436	0.372	0.451	0.391	0.450	9
	11	0.399	0.416	0.462	0.415	0.388	0.444	0.386	0.460	0.406	0.460	8
	12	0.425	0.447	0.501	0.446	0.414	0.480	0.413	0.500	0.438	0.500	7

TABLE 3.15
Overall Score and Ranking (Professionals)

Type of Industry	Ind. No.	Overall Score and Combination of Mass										Rank
		1,2	1,3	1,4	1,5	2,3	2,4	2,5	3,4	3,5	4,5	
Chemical	1	0.811	0.804	0.804	0.805	0.805	0.804	0.805	0.795	0.795	0.795	2
	2	0.829	0.829	0.829	0.829	0.830	0.830	0.830	0.830	0.830	0.830	1
	3	0.769	0.745	0.745	0.744	0.744	0.744	0.744	0.711	0.710	0.711	3
Textile	4	0.564	0.569	0.569	0.569	0.544	0.543	0.544	0.545	0.546	0.544	5
	5	0.544	0.555	0.554	0.553	0.528	0.528	0.527	0.536	0.535	0.534	6
	6	0.570	0.579	0.579	0.578	0.549	0.549	0.548	0.554	0.554	0.552	4
Engg. (Automobile)	7	0.296	0.286	0.286	0.286	0.266	0.266	0.266	0.255	0.254	0.254	12
	8	0.298	0.291	0.291	0.291	0.269	0.269	0.268	0.261	0.260	0.261	11
	9	0.317	0.323	0.323	0.322	0.286	0.287	0.286	0.290	0.289	0.290	10
Dairy	10	0.380	0.379	0.381	0.379	0.360	0.363	0.360	0.361	0.358	0.362	9
	11	0.392	0.393	0.395	0.393	0.373	0.375	0.373	0.375	0.373	0.376	8
	12	0.417	0.421	0.422	0.420	0.395	0.397	0.395	0.399	0.397	0.399	7

REFERENCES

Alfassi, Z. B., Boger, Z., and Ronen, Y. (2004). *Confidence Limits of the Mean, Statistical Treatment of Analytical Data*, Wiley, Blackwell Science Ltd.

Ayyub, B. M. (2003). *Fundamentals of Probability and Statistics, Risk Analysis in Engineering and Economics* Second Edition, Chapman & Hall.

Ayyub, B. M. (2014). *Fundamentals of Probability and Statistics, Risk Analysis in Engineering and Economics* Second Edition, Chapman & Hall.

Black, M. (1937). Vagueness: an exercise in logical analysis. *Philosophy of Science*, 4(4), 427–455. (Reprinted in International Journal of General Systems, 17 (2-3)), 1990, 133–141.

Buckley, J. J., and Hayashi, Y. (1994). Fuzzy neural networks. *A survey. Fuzzy Sets and Systems*, 66, 1–13

Chen, S., and Hwang, C. (1992). *Fuzzy Multiple Attribute Decision Making- Methods and Applications*, Springer-Verlag, Berlin, Heidelberg, New York.

Dodgson, J. M., Spackman, A. Pearman, and Phillips, L. (2001). *Multi Criteria Analysis: A Manual*, Department for Transport, Local Government and the Regions, London.

Edwards, W. (1977). Use of multiattribute utility measurement for social decision making. In: D.E. Bell, R.L. Keeney and H Raiffa, Editors, Decisions John Wiley and Sons, Chichester, 247–276.

Emery A. Coppola Jr, Lucien Duckstein, and Donald Davis (2002). Fuzzy rule based methodology for estimating monthly groundwater recharge in a temperate watershed. *Journal of Hydrologic Engineering*, 7, ASCE 1084-0699,7:4 (326).

Hipel, K. W., Radford, K. J., and Fang, L. (1993). Multiple participant–multiple criteria decision making. *IEEE Transactions on System, Man, Cybernetics*, 23, 1184–1189.

Hwang, C.L., and Yoon, K. (1981). *Multiple Attribute Decision Making: Methods and Applications*, Springer Verlag, New York.

Kaufmann, A., and Gupta, M.M. (1991). *Introduction to Fuzzy Arithmetic Theory and Application*, Van Nostrand Reinhold, New York.

Klir, G. J., and Yuan, B. (2003). *Fuzzy Sets and Fuzzy Logic, Theory and Applications*, Prentice Hall of India Private Limited, New Delhi.

Lad, R. K. and Christian, R. A. (2010). A fuzzy framework for environmental pollution potential and acceptability of industries in a region, Ph.D. Thesis.

Lad, R. K., Desai, N. G., Christian, R. A., and Deshpande, A. W. (2008). Fuzzy modeling for environmental pollution potential ranking of industries. *Environmental Progress*, 27(1), 84–90

Malczewski, J. (1999). *GIS and Multicriteria Decision Analysis*, Wiley, New York.

Saaty, T. L. (1980). *The Analytic Hierarchy Process*, McGraw Hill, New York.

Saaty, T. L. (1977). A scaling method for priorities in hierarchical structures. *Journal of Mathematical Psychology*, 15, 234–281.

Sattar, J Aboud (2009). Soft Computing Techniques for Protecting Information Security Systems. *MASAUM Journal of Computing* 464-467, 1(3).

Spiegel, M. R., and Stephens, L. J. (1999). *Schaum's Outline of Theory and Problems of Statistics*. McGraw Hill, Singapore.

Warren McCulloch and Walter Pitts (1943). A logical calculus of the ideas immanent in nervous activity. *Bulletin of Mathematical Biophysics*, 5, 115–133.

Yousef Sajed, Gholamali Shafabakhsh, and Morteza Bagheri (2019). Hotspot location identification using accident data, traffic and geometric characteristics. *Engineering Journal*, 23(6). pp 191–207.

Zadeh, L. A. (1965). Fuzzy sets. *Information and Control*, 8(3), 338–353.

4 Structural Engineering: Trends, Applications, and Advances

Deepa A. Joshi and Manisha Shewale

4.1 OVERVIEW OF STRUCTURAL ENGINEERING

Structural engineering is one of the most important fields in civil engineering. Structural engineering is basically the field that is responsible for making the structure stand safe for its entire life span. The word 'structure' here implies any type of structure, such as buildings, bridges, water-retaining structures, retaining walls, dams, etc. Reinforced cement concrete and steel structures are the most common types of structures. Structural engineering involves analysis, design, monitoring, damage detection, repair, rehabilitation or retrofitting, and preservation of historical monuments. Basic structural engineering deals with the analysis and design of the structural members. The members carrying load are termed as 'structural members' in the structure; for example, in load bearing structures, walls carry the load and hence walls are the structural members, whereas, in framed structures like buildings, the structural members are slabs, beams, columns, and foundations. The function of structural members is to carry the load safely and transfer it to another structural member.

The loads are mainly divided as follows:

- Gravity Loads
  - Dead Load
  - Live Load
- Lateral Loads
  - Wind Load
  - Earthquake Load

Gravity load means the load that acts under gravity, vertically downward, and the lateral load means the loads act in lateral directions on the members. The gravity loads consist of dead load and live load. Dead load implies the weight of members such as weight of slab, beams, walls, columns and foundations. A dead load is estimated from the weight density of material. The live load includes the load of users, furniture, equipment, etc. A live load depends on the functional use of the structure, such as live load for a small bungalow, for an education complex, for a mall will differ and hence live load is taken from IS 875 Part 2 [1]. Standard

DOI: 10.1201/9781003316671-4

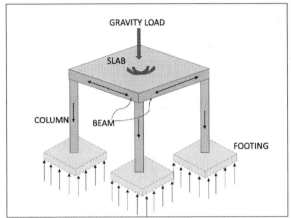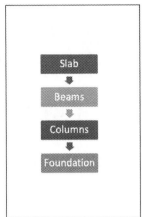

FIGURE 4.1 Load Transfer Path in Building.

procedures for determination of wind load and earthquake load are provided by the respective Indian Standard codes (IS Codes). Also, in some cases, an additional load such as snow load in the region, where snow gets accumulated on roof tops of structures, needs to be considered.

Hence, it is very much essential to understand the load transfer path and calculate accurate loads, for safe design of any structure. In the case of buildings, the load gets transferred from the slab to beam, beam to column, column to foundation, and finally from foundation to the soil strata. The load transfer path in a building structure, as shown in Figure 4.1.

In the case of bridges, the load of bridge decks is transferred to bridge girders. The girders transfer the load to bridge piers and further it is transferred to the foundation. If we consider the dam structure, then the water pressure is taken by the huge dam wall. Hence, in every type of structure, it is important to identify the load transfer path, the structural members, and finally design all the structural members.

The task of design for new construction can be divided into two main tasks: analysis and design (Figure 4.2).

- **Structural Analysis:**
  Structural analysis is a prime important task in structural design and it needs thorough understanding of the behavior of the structure/structural members. The structural members are subjected to various loads throughout the life span of the structure, as discussed above, and hence various types of stresses and deformations are developed in the members. Structural analysis includes determination of all the loads coming on the structural members and their effect accurately. The accuracy in structural

FIGURE 4.2 Major Task in Structural Design.

FIGURE 4.3 Major Parameters in Structural Analysis.

analysis is very important as members are designed on the basis of analysis results. There are various methods, techniques, problems, and challenges in structural analysis as structural analysis itself is a very vast field.

'Structural analysis' can be defined as estimation of the loads and effects of loads on structural members, with understanding of their structural behavior. The major parameters to be determined in structural analysis are presented in Figure 4.3.

- **Structural Design:**
  Structural design is carried out after completion of the detailed and accurate structural analysis. The structural design basically involves providing the dimensions and reinforcement in required form, so that the structural member will be able to resist the load/stresses and will stand throughout the life span of the structures. The structural design depends on two major aspects: analysis results and the type of structure/material.

  On the basis of material, various types of structures are
  - Masonry structures
  - Reinforced cement concrete (RCC)
  - Steel structures
  - Prestressed concrete structures

The structural design also depends on the construction methods like cast-in-situ or precast. Structural design must be carried out as per the provisions in IS Codes. IS Codes are the Indian Standard Codes for the design of structures must be followed in India. Other countries have their own codes, such as ACI (American Concrete Codes), BS codes (British codes), and so on.

Finally, the structural drawings are prepared on the basis of the structural design and these drawings are provided on the construction site. The construction is carried out exactly as per the structural drawings. Hence, it is the responsibility of structural engineer to provide a safe and optimum design for the structure.

Figure 4.2 provides tasks for new construction; however, for every structure, one major task is necessary after construction and that is 'maintenance'. *Maintenance* is the term that generally covers regular maintenance of the structure and repair if needed. Maintenance largely depends on the type of structure, as there is a huge difference in maintenance of a residential building and a dam. Structural audit, structural health monitoring, retrofitting of structures, preservation of historical monuments, etc. are the advanced fields under structural engineering.

4.2   NEED OF DATA SCIENCE IN STRUCTURAL ENGINEERING

As discussed in section 4.1, structural engineering involves analysis, design, monitoring, earthquake engineering, damage detection, repair, rehabilitation or retrofitting, preservation of historical monuments, prediction of concrete properties, and development of new materials. Generally, analysis and design of regular structures is carried out with the established methods and IS Codes. Structures that involve complexity in terms of geometry, loading, site conditions or any other aspect, need experienced experts. Also, the retrofitting or strengthening projects are unique in nature; that means every project is different and needs specific analysis and design. The development of new structural materials or new techniques of utilizing various materials for structural members involves a huge amount of research and especially experimental work.

Major reasons for necessity of application of data science techniques in structural engineering are discussed in brief here.

1. **Application of data science techniques can minimize huge experimental work involved in structural engineering, thereby saving time, cost, and labour.**

Experimental work is an essential part of structural engineering. Experimental work under structural engineering can be broadly classified as 'routine testing' and 'experimental work for R&D'.

Routine Testing: Routine testing is necessary even for a small building project, and use of conventional material and conventional techniques, then also testing of concrete is necessary. The test results of compressive strength of concrete after 28 days are considered for full-strength results. Hence, in every project, testing is one of the essential tasks. If we consider big projects or important projects such as bridges, dams, thermal power plants, nuclear power plants, etc., one can imagine the experimentation needed for such projects.

By using data science techniques, the properties of material can be predicted. These advanced techniques need few parameters as input and the required specific property can be obtained as output. This can reduce huge amount of routine experimental work in the construction industry.

Experimental work for R&D: Apart from routine experimental work, the experimentation is an integral part of development of new material or new structural technology. R&D work in structural engineering is huge in quantity as well as in nature, such as testing of large-scale columns, beams, beam-column junctions, large-scale walls, and even entire buildings. The experimental work for R&D needs detailed and systematic planning, exhaustive preparation, and experts' monitoring. The equipment, instrumentation, and supporting structures needed for such work are huge. Hence the experimental work involved in R&D under structural engineering is very complicated.

In general, the structural engineering involves huge experimental work and application of data science techniques has the potential to minimize the experimental work, through which a significant amount of savings in terms of time, efforts, labour, and cost can be achieved.

2. Real-time behavior of the structural members can be studied and maintenance work can be planned.

Civil engineering structures are large-scale structures and, once constructed, these are supposed to resist all the loads and give safe service in the entire design life span. Earlier, it was difficult to find out exact condition of the structural members and know whether any kind of distress or excessive deflection or cracking has taken place which may lead to sever damage to the member or entire structure, unless and until some major visible damage is observed. In such a situation, the entire process of identifying the failed members, repairing, or strengthening becomes a very complicated and costly affair. Instead, if the behavior of structural members can be monitored regularly and any kind of defect can be detected at early stage, it can be easily rectified without much damage to the entire structure. This kind of real-time monitoring is possible only because of the advancements in the data science. This need of real-time monitoring and damage detection of structures is the main cause of development of advanced field under structural engineering, termed as 'structural health monitoring' (SHM). As the name indicates, it involves monitoring the health of the structures on the basis of data collection through various types of sensors. The collected data is continuously monitored, observed and studied. Whenever, any kind of irregularities are observed, the causes of such irregularities are found out and required structural maintenance can be suggested. For major important projects such as bridges, dams, nuclear power plants, etc. SHM is carried out.

3. Applications of data science techniques are very useful in avoiding big accidents such as collapse of structures.

The majority of structures weaken as they age, strained not only by the loading but also due to the effects of extreme cold, water, and salt corrosion. Earthquakes and flooding can also cause destabilization of the structures, and such weaknesses can cause catastrophic failure if unnoticed.

There are number of incidences in which the entire bridge has collapsed or multistoried building has collapsed, which has resulted in death of many innocent people. In 2016, strong rains caused the collapse of a bridge built in the British era over the Savitri River on the Mumbai-Goa Highway in Mahad in the Raigad district of the Konkan region. Two state transportation buses, in which passengers were traveling, were washed away as a result of this occurrence. In the Indian city of Mumbai, a portion of a pedestrian bridge connecting the north-end of the Chhatrapati Shivaji Maharaj Terminus (CSMT) train station to Badaruddin Tayabji Lane broke and fell onto the road in 2019. The catastrophe claimed the lives of six people and injured at least another 30 others [2]. In 2020, the Bihar bridge collapsed; no injuries have been reported in this incident, despite the fact that there were no casualties recorded in the incident, the bridge was built at a cost of Rs 263.48 crore, according to the state government of Bihar [3]. Again in 2020, an under-construction bridge in West Bengal's Malda collapsed leaving at least two dead and five others injured, and a huge economic loss [4].

As discussed in an earlier section, the SHM is very effective in early damage detection in the structural members. This definitely reduces the maintenance cost; however, the major benefit of SHM in big accidents such as collapse of structures can be completely avoided and human life can be saved. Hence the SHM, which is possible because of advancements in sensors, data collection, pattern recognition, data analysis, etc., is an effective technique under structural engineering, by which collapse of structures can be prevented and further huge losses in terms of human life and money can be avoided.

4. Structures in remote areas can also be monitored.

As discussed in the points above, maintenance and monitoring of structures through conventional methods have many challenges that have been overcome with the help of data science. Data science proves helpful especially for the structures that are located in the remote areas like the structures constructed in hilly regions and the underwater structures where monitoring using the conventional methods is not feasible.

This can be well explained with the help of an example of an ongoing 8 lane, 22 kms coastal project in Mumbai's western coastline connecting Marine Lines in the south to Kandivali in the north. A big challenge lies here in monitoring the structure. Data science applications like the use of remote operated vehicles (ROVs) can be used to leverage streaming data by setting up a monitoring dashboard by logging statistical profiles of the streamed data. Faults can also be monitored with the help of a low-cost telerobotic underwater drone. The monitoring dashboard can serve two purposes: to monitor the performance and quality issues of the regression model, but also (and mostly) to monitor the health of the vehicle itself. The breakdown of a ROV can again be monitored by using data science tools i.e., manually injecting some sensor faults in the data to detect faults through the use of our dashboard.

Thus, applications of data sciences prove very helpful for overcoming the problem of monitoring of remote structure which is the essential need of an hour.

5. Development of new materials or techniques.

Development of new materials or design philosophies, construction techniques are an integral and essential part of R&D in structural engineering. The evolution of the construction sector shows that initially, with the availability of only stones and bricks as construction materials, only masonry structures were constructed untill cement was invented and were considered binding material, giving rise to the development of concrete. Concrete changed the entire picture and framed structures came into existence. The framed structures are still in practice and new technology of using the same two important materials, concrete and steel, with different philosophies was invented as prestressed concrete. Hence, it can be seen that either new materials or new technologies are continuously improving the field of structural engineering.

Along with the developments in structural engineering, the data science has also developed rapidly and the interdisciplinary approach of applying the data science techniques in structural engineering gave a boost in developments of new materials and techniques. The prediction of properties of the martials by developing the analytical models, monitoring the stress-strain behaviour of members, using pattern recognition techniques, etc., have been proved to be very effective and accurate options for establishing new material or techniques. Hence, we can conclude that the advancements in data science have geared the speed of technological advancements in the field of structural engineering.

6. Time saving.

'Time' is a crucial parameter when it comes to design, construction, maintenance, repair, and retrofitting of structures. The advancements in data sciences are being applied intelligently in structural engineering, which definitely led to 'time savings'. Whatever points are discussed above, starting from minimization of experimentation, real-time monitoring of structures, timely maintenance, prevention of collapse of structures,and development of new material or technology, each has a positive impact on time management. Considering the vastness of the construction industry, this 'time savings' due to application of data science techniques definitely has a massive and positive effect on the economy of the nation.

7. Cost saving.

Finally, 'cost' is the most important and critical factor in any kind of project and it needs no explanation. Whatever points are discussed above, starting from minimization of experimentation, real-time monitoring of structures, timely maintenance, prevention of collapse of structures, development of new material or technology, and including 'time saving'; each has a huge positive impact on the cost/economy of the project.

Here, one point is to be noted that, the application of data science techniques in structural engineering definitely saves the cost; however, at the same time, it also demands the additional cost accounting to implementation of these techniques. Therefore, a cost-efficient technique needs to be selected from all the available options and it definitely differs from project to project.

Structural engineering itself is a very vast field and application of advanced techniques of data science is possible in many problems under it, such as structural design optimization, prediction of various properties of various types of concretes, damage detection, prediction of strengths of retrofitted structural members, fire resistance evaluation, structural health monitoring, etc.

Application of data analysis not only give accurates and real-time results but also can minimize the huge experimental work and can prevent the collapse of structures, thereby saving human life, can gear up the developments in terms of new materials or technologies in structural engineering and hence will be cost effective and time saving, optimizing the materials and human resources.

4.3   CURRENT TRENDS AND APPLICATIONS OF DATA SCIENCE IN STRUCTURAL ENGINEERING

Application of data analysis and the advanced techniques of AI in structural engineering is gaining popularity due to its various advantages. Structural engineering itself is a very vast field and application of advanced techniques of data analysis is possible in many problems under structural engineering. In this section, various data analysis techniques with a huge scope in structural engineering problems are presented. The objective of including this section in this chapter is to make the readers aware up to what extent the work in this field has been carried out. An attempt has been made to discussa majority of applications.

4.3.1   GENETIC ALGORITHM

The genetic algorithm (GA) is a well-known evolutionary algorithm proposed in 1969 by University of Michigan Professor Holland and detailed by Holland et al. [5–7]. GA is a form of random algorithm inspired by Darwin's theory of evolution that simulates the natural process of biological evolution [8]. GA is a technology that uses natural selection and survival of the fittest to identify the best answer. This section goes over a few of the applications of GA in structural engineering.

Structural health monitoring (SHM) is a rapidly growing discipline that necessitates the use of modern and cutting-edge technologies. SHM is becoming more appealing as a result of its capability for damage detection. Silva et al. (2016) [9] proposed an unstructured and unregulated genetic algorithm for decision boundary analysis (GADBA) to aid in structural degradation identification and applied the method to the Tamar Bridge in the United Kingdom.

As a result of the sluggish convergence rate of genetic algorithms (GAs) in large-scale optimization, Mroginski et al. [10] presented a selective genetic algorithm (SGA) for the optimization of truss cross-sections. To minimize the search space for the evolutionary process, SGA is based on a primary sensitivity analysis of each variable performed on each variable.

Truong et al. [11] employed practical advanced analysis (PAA) and a micro-genetic algorithm (GA) to optimize space steel frames with semi-rigid joints, using cross-sectional areas of beams, column members, and the types of semi-rigid connections as optimization factors. Artar et al. [12] employed a genetic algorithm and a harmony search method to improve the design of steel frames with semi-rigid connections.

Zarbaf et al. [13] devised a genetic algorithm (GA) and particle swarm optimization (PSO) technique for calculating stay cable tensions of cable-stayed bridges, taking into account the influence of bending stiffness and vertical elongation.

Designed to lessen structural responses while simultaneously protecting sensitive internal hardware and non-structural parts, seismic base isolations are nearly ideal passive control devices that limit structural responses. Mehrkian et al. [14] proposed a multi-objective fuzzy-genetic control system for structural vibration reduction in an irregular base-isolated benchmark building under a variety of seismic situations.

The evolutionary algorithm is commonly utilized in civil engineering because it provides a general framework for tackling large-scale system optimization problems that is independent of the specialized discipline in which the problem is being addressed. However, despite the fact that the evolutionary algorithm has yielded great results in actual engineering, it still has several drawbacks. Because of this, genetic algorithm academics from all around the world have been working on enhancing and developing their algorithms.

4.3.2 Swarm Intelligence

In 1992, Hackwood et al. [15] proposed the concept of swarm intelligence (SI) in the molecular automata system, which achieves self-organization through inter-connections between neighbours in the grid space. SI was first used to describe algorithms based on the collective behaviour of social insects and other animals [Bonabeau, Dorigo, and Theraulaz (1999)] [16]. "Research into the collective be-haviour of multi-component systems that are coordinated by decentralised controls and self-organization," Saleem, DiCaro, and Farooq (2011) [17] describe SI. Swarm intelligence–based meta-heuristics simulate a group of simple individuals, evolve their solutions through interactions and interactions with the environment, and show good performance on a variety of difficult problems and have thus become a very active research field in recent years [18].

4.3.2.1 Particle Swarm Optimization

Several researchers, including Kennedy et al. [19], have proposed particle swarm optimization (PSO) as a method for simulating the behavioral patterns of animals or insects (such as birds and fish). Particle swarm optimization (PSO) is a population-based global optimization technique that allows for the movement of many independent solutions, referred to as particles, around a hyperdimensional search space in pursuit of the ideal solution.

The assessment of surrounding rock deformation during underground excavation can aid in the minimization of possible risk and the selection of design parameters for further development and development of underground excavation. However, because there are a variety of factors influencing the distortion of surrounding rocks in underground caves, it is difficult to establish their relative importance in terms of deformation of surrounding rocks. Nonetheless, in order to overcome this issue, Xue et al. [20] introduced a least squares support vector machine (LSSVM) technique based on the particle swarm optimization (PSO) algorithm for analyzing the deformation of surrounding rocks in underground caves.

Given that the shear wall of a squat reinforced concrete (RC) building with a low aspect ratio is critical for resisting lateral seismic loading, developing a prediction model for its shear capacity is necessary for insuring the seismic safety of buildings in seismically active areas. When Chen et al. [21] wanted to predict the shear strength of squat reinforced concrete walls, they employed a hybrid technique that combined an artificial neural network and particle swarm optimization, which they invented themselves (ANN-PSO).

Qin et al. [22] used the Kriging model and particle swarm optimization (PSO) to update the dynamic model of Spain's Jalón Viaduct using higher vibration modes under initial conditions of large vibration amplitude in order to extract higher modes and improve computational efficiency when updating dynamic models of bridge structures.

The Kriging model is used to forecast analytical answers as a surrogate model for a sophisticated finite element model, while the PSO method is utilized to update the original finite element model.

In comparison to genetic and ant colony algorithms, the PSO algorithm is simple, requires fewer adjustment parameters, has a fast convergence rate, and has a high degree of robustness, which is why it is widely employed in a variety of applications, including structural design, functional optimization, pattern categorization, and fuzzy system control, among others.

4.3.2.2 Bee Colony Algorithm

In 2005, Karaboga [23] proposed the artificial bee colony approach, which was afterwards used. In general, bees cooperate with one another to carry out honey-gathering chores by splitting labour and sharing knowledge amongst themselves, which is the central tenet of the book. However, even if individual bees have limited capabilities, it is still simpler for the entire colony to select high-quality honey sources without requiring the same training as an individual bee. Instead of relying on objective function and constraint as in standard optimization methods, this algorithm relies just on fitness function for evolution, which is in stark contrast to traditional approaches.

It is challenging to develop an appropriate response surface for traditional slope stability analysis due to the complexity of soil parameters. In Kang et al. [24], intelligent response surface technique for system probabilistic stability evaluation of soil slopes, support vector regression (SVR) was refined using the artificial bee colony algorithm (ABC) to generate the response surface and approximate the limit state function.

Yang [25] introduced the firefly algorithm, a heuristic algorithm inspired by the flashing activities of fireflies that uses nodes in search area to replicate individual fireflies in nature. Individual fireflies are used to simulate the attraction and movement of the search. Individual's position is used to measure the objective function of solving the challenge. In the search and optimization process, the individual survival of the fittest process is equivalent to the iterative process in which the better feasible solution replaces the inferior option [26].

At the moment, the approaches for optimizing the shape and size of multi-frequency limited truss structures are difficult to design and put into practice because of the time commitment required. Lieu et al. [27] introduced an adaptive hybrid evolutionary firefly algorithm (AHEFA) that is based on the differential evolution (DE) algorithm and the firefly method, with the goal of improving the algorithm's convergence speed and accuracy. A combination of the DE algorithm and the firefly method was used to develop the algorithm (FA). It is possible to select the most appropriate mutation approach for the trade-off between global and local search capabilities through the use of adaptive parameters, which may be found in the literature.

When it comes to updating experimental models of structures, the technique of numerical model updating is applied. Nevertheless, these updating methods still have some restrictions and ambiguities, which makes it difficult to comprehend the structures they are used in. A numerical model, designed by Kubair et al. [28] and based on the firefly technique, was constructed to efficiently reduce the gap between experimental and analytical data in frame and cantilever beam instances.

FA offers several advantages over other algorithms in terms of optimization speed and accuracy, despite its limited development history. However, significant issues with its application remain, such as parameter selection, premature convergence, and a shaky theoretical foundation.

4.3.2.3   Cuckoo Search

Swarm intelligence technology was used to develop the groundbreaking nature-based heuristic algorithm known as cuckoo search (CS), which was first developed by British academics Yang et al. in 2009 [29–31] and then refined by other researchers. Observations of cuckoos' nesting parasite activities and bird Lévy flight behaviour were used to develop the algorithm's notion. Because of a variety of competing design objectives, it is difficult to solve the problem of semi-active control of intelligent isolated buildings under near-field and far-field earthquake excitations. Using a multi-objective cuckoo search algorithm to optimise the parameters of a fractional order proportional-integral-derivative (FOPID) controller, Zamani et al. [32] were able to meet their control objectives while also reducing deformation of the isolation system without significantly increasing acceleration of the superstructure. In engineering optimization, the Cuckoo search (CS) approach has a great deal of potential because it is straightforward to implement and requires just a small number of parameters. An intelligent multiple search strategy algorithm (IMSS) was introduced by Rakhshani et al. [33] that uses covariance matrix adaptation evolution strategy (CMAES) and covariance matrix adaptation evolution strategy (CS) to effectively explore and reduce the computing time required to find the best answer.

4.3.3   ARTIFICIAL NEURAL NETWORKS

Artificial neural networks (ANNs) are a type of system used in data processing and knowledge representation that is comprised of tightly interconnected adaptive simple processing elements (referred to as artificial neurons or nodes) that are capable of performing large-scale parallel computations [34,35]. It was that created the world's first artificial neuron model (the MP model), which employed a reduced signal propagation mechanism to mimic some of the fundamental processes of human brain neurons [36]. This concept served as a foundation for the early development of brain computing, which was pioneered by IBM in the 1960s and has since become widely used.

Among the first journal articles on neural network implementations in civil and structural engineering were those by Adeli and Yeh (1989) [37]. Since then, neural networks have been increasingly popular in the field of civil engineering. ANNs are a technological reconstruction of biological neural networks, and in this sense, they

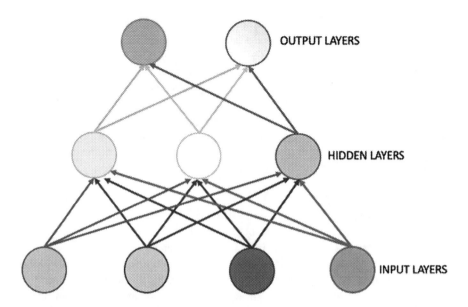

FIGURE 4.4   Three-Layer Feedforward Network.

are similar to artificial neural networks. In particular, it seeks to develop a realistic artificial neural network model based on the principles of biological neural networks and the requirement for practical application, as well as to develop a learning algorithm for it and to duplicate some intelligent human brain functions. Last but not least, it is put into action on a technological level to deal with real-world problems. According to Basheer and Hajmeer (2000) [35] and Hecht-Nielsen (1990) [38], Figure 4.4 shows three-layered feedforward network.

A few applications of artificial neural networks in the field of structural engineering are addressed briefly below.

Surface settlement is one of the most harmful variables in infrastructure tunnel excavations, and hence accurate surface maximum settlement (MSS) prediction is essential for reducing the risk of surface failure in these tunnels. A hybrid artificial neural network (ANN) model based on particle swarm optimization (PSO) was proposed by Hasanipanah et al. (2016) to predict MSS induced by tunnelling along Karaj subway line 2 [39].

Bridges should be built with safety and long-term durability in mind in order to save overall costs. To reduce the computing time of the finite element model used in deck analysis, Garcia-Segura et al. [40] developed a multi-objective harmony search with artificial neural networks (ANNs) using artificial neural networks (ANNs). The best design of a post-tensioned concrete box girder road bridge takes into consideration the cost, overall safety factor, and early corrosion period, among other considerations.

Fiber reinforced polymers (FRP) are widely utilized in passive confinement to reinforced concrete to increase the compressive strength and ductility of the concrete (RC). A model for forecasting fiber-reinforced concrete compressive strength based on an artificial neural network was proposed by Casardi et al. [41] and was based on an artificial neural network (ANN). In contrast to other models, the one proposed creates an analytical relationship without taking into consideration the conventional efficacy parameter.

In comparison to reinforced concrete products, glass fiber–reinforced polymer (GFRP) bar reinforced–concrete constructions have greater durability. When it comes to the design and application of polymer-matrix composites, it is critical to evaluate the bonding qualities of fibreglass reinforced plastic (GFRP) bars to concrete (PMCs). Chinese researchers Yan et al. [42] developed a hybrid model to predict the binding strength of glass fibre reinforced plastic bars to concrete by merging the strong non-linear mapping capacity of an artificial neural network with the global searching ability of a genetic algorithm (GA) [42].

Because incorrect structural design can lead multistory reinforced concrete structures to fail prematurely, it is necessary to conduct a thorough assessment of all elements affecting the structure before proceeding with the project. NN-PSO is a model developed by Chatterjee and colleagues [43] for predicting the structural failure of a multistory reinforced concrete building. This is because the conventional neural network model has poor convergence under the training of the local search optimization algorithm, and thus does not produce the expected learning effect.

A variety of approaches have been developed in recent years to forecast and analyze the properties of recycled aggregate concrete (RAC), which has the potential to significantly minimize construction waste while maintaining structural integrity. An ANN model trained on 139 sets of available data was utilized by Naderpour et al. [44] to predict the compressive strength of RAC. The water-cement ratio, the water absorption rate, the fine aggregate, the natural coarse aggregate, the recycled coarse aggregate, the water-total material ratio, and the water-total material ratio are all taken into consideration by the proposed ANN model, which has six input parameters.

The ANNs perform admirably in some areas, while failing badly in others. These algorithms are particularly well adapted to professions that need the use of partial data sets, ambiguous or missing information, and highly difficult and ill-defined circumstances, in which human decisions are frequently made intuitively [Kalogirou (2001)] [45] and Adeli (2001) [46] explains how neural networks have been widely used in civil engineering for decades in applications such as structural optimization, structural condition assessment and health monitoring, structural control, structural material characterization and modeling, construction engineering, highway engineering, and other fields. With notable successes in speech recognition, image recognition, natural language processing, and other domains, deep learning has emerged as a popular research topic in neural network research in recent years, gaining popularity among researchers.

4.3.4 BIG DATA

Big data approaches have rapidly advanced in recent years, becoming a popular issue in scientific and technology fields, commerce, and even administrative institutions around the world [47]. Nature produced a special version on big data in 2008, which may be found here. In February 2011, *Science* published a special issue titled "Dealing with Data", which was devoted to the subject of big data in science investigation and included examples of how large data may be beneficial in academic studies. In the literature, big data is defined by three components: I volume (terabytes, petabytes, and beyond); (ii) variety (structured and unstructured data); and (iii) velocity (terabytes, petabytes, and beyond). Big data is defined as follows: continuous streams of the data. It is necessary to adapt data from a variety of new sources in engineering [48], such as GPS, wireless devices, sensors, and streaming communication generated by machine-to-machine interactions. When dealing with data that is continuous, unstructured, and not restricted by the rigid structure of rows and columns, traditional techniques have difficulty coping. Big data applications are therefore crucial in this context.

Massive data sets with distributed and decentralized governance distinguish big data applications from other types of data processing. The fact that each data source is self-sufficient means that it can produce and gather data without any need for (or reliance on) centralized control [49]. Similar to the World Wide Web arrangement, in which each web server provides a set volume of data and is totally functional without the assistance of other servers, this is similar to the arrangement used by the Internet. For those systems that are entirely reliant on a single centralized control unit, the sheer volume of data may make programmes vulnerable to attacks or malfunctions due to the sheer volume of data. To provide seamless services and quick response for local markets, a number of server farms are strategically deployed throughout the world for significant big-data-related apps such as Google and Facebook. This source of autonomy is the result of a combination of national and regional law and policies, as well as a technological solution to the problem.

4.4 APPLICATION OF AI IN CONCRETE TECHNOLOGY

'Concrete technology' can be considered one of the prime areas of structural engineering or civil engineering. Concrete technology is all about 'concrete'. Concrete is a high-demand material in the construction industry and plays a vital role in development. Concrete is used twice as much as all other building materials combined, including wood, steel, plastic, and aluminum, in construction all over the world [50]. Concrete is the second most utilized material after water.

Massive amounts of research are being carried out all around the world in order to manufacture concrete:

- more and more effective,
- strong,
- economical,
- minimizing adverse effect on environment,
- more sustainable.

The recent trend is to utilize AI techniques for such research work, which has major advantages over conventional methods.

In this section of the chapter, application of AI in concrete technology has been discussed in detail. The first part of section focuses on 'concrete', in which all basic information related to concrete has been provided. Further, the applications of AI are discussed for two major tasks related to concrete; first, for predication of mechanical properties of concrete and second, mix design of concrete. Hence, this section will take the readers smoothly from basics of concrete to application of AI in concrete technology.

4.4.1 CONCRETE

Concrete is the basic and most important material for construction industry and the concrete technology is one of the most important field for civil engineers or structural engineers. Concrete is used in buildings, bridges, roads, dams, and almost in each and every man-made structure. This makes concrete the world's largest used construction material. The work in the field of infrastructure is ever increasing in developed as well as developing countries and hence the demand of concrete is also increasing day by day.

Concrete is basically a heterogeneous material and the basic ingredients of concrete are

- Cement
- Fine Aggregate (sand)
- Coarse Aggregate
- Water

Each of the ingredients has a specific role in concrete.

- *Cement*: It is the binding material. It provides strength to the concrete. Because of its setting and hardening capabilities, it is able to bind the aggregates together to form a solid mass. Cement also plays a role of filling the voids in fine aggregates, thereby making the concrete impermeable. Ordinary Portland cement (OPC) is the most common type of cement, and used all over the world. The name 'Portland cement' is derived from its resemblance to 'Portland stone', which was quarried in England. It was developed in England in the early nineteenth century. As of today, there are many types of cements developed and being utilized in the construction industry, serving a specific purpose. Following are some of the frequently used types of cement:
  - Ordinary Portland Cement (OPC)
  - Rapid Hardening Cement
  - Low Heat Cement
  - Portland Pozzolana Cement
  - Quick Setting Cement
  - White Cement
- *Aggregates*: Aggregates are a significant component of concrete, accounting for 70 to 80% of the total volume of the material. The characteristics of

concrete are significantly influenced by the aggregates used. Aggregates are critical for the strength and thermal and elastic properties of concrete, as well as for dimensional stability and volume stability of the finished product. The use of aggregates in concrete helps to prevent shrinkage and cracking while maintaining structural integrity. The two forms of aggregates are coarse aggregates and fine aggregates. Fine aggregates are the most common type of aggregate. Crushed stone is comprised of coarse aggregates with sizes higher than 4.75 mm in diameter, while fine aggregates, such as sand, have sizes smaller than 4.75 mm in diameter.

- *Water*: Water is a crucial component of concrete production. The adhesive property of cement, which is a binding material, is only obtained when it has been mixed with water. Hydration of cement is the phrase used to describe the chemical interaction that occurs between cement and water. This reaction is exothermic, which means that it generates heat, which is referred to as the 'heat of hydration.' It is critical that the correct amount of water is added to the mixture. Regarding the quality of the water, the IS standards specify that only potable water should be used for concrete production.

Along with this these basic ingredients, various admixtures are added with specific purpose such as to produce high performance concrete, high strength concrete, self-compacting concrete, etc.

Not only the type and quality of the ingredient is important but the quantities of these materials are also important. The quantities of these materials have an effect on the properties of concrete. The proportion of ingredients is decided by the mix design.

Concrete is recognized by its grade, such as M20, M25, M30, etc., where 'M' stands for 'mix' and the number 20, 25, etc. is the compressive strength of the concrete. Compressive strength is the most important property of concrete. The concrete is known by its compressive strength. Following are the important properties of concrete:

1. *Compressive Strength*: It is the capacity of concrete to resist compressive stresses. It is determined in a laboratory by applying compressive load on 150 mm concrete cube after curing for 28 days. The compressive strength is expressed in N/mm^2 or MPa.

2. *Characteristic Strength of Concrete*: This is defined as the value of compressive strength of concrete, below which not more that 5% of the test results are expected to fall. This means there is only 5% chances of actual strength being less that designed strength. This value is denoted by 'f_{ck}'.

3. *Tensile Strength of Concrete*: Tensile strength of concrete is very low. The concrete resists compressive stresses; however, it is very weak in tension. Approximately the tensile strength of concrete is hardly 10% of the compressive strength. As per the IS code (IS 456: 2000 for RC Design) [51], the tensile strength of the concrete is neglected and hence not considered in the design of reinforced concrete (RC) structures.

Apart from the above properties, workability is one of the important properties of fresh concrete and creep, shrinkage, and durability are the properties that define long-term behavior of the concrete. Modulus of elasticity and Poisson's ratio are essential for material characterization of concrete and these properties are required for the simulation/analysis/design by software.

A large amount of work is being carried out in the field of concrete technology and various attempts are being made to enhance the effectiveness of concrete and reduce the harmful effects of ingredients of concrete on environment. This is achieved either by adding admixtures or other materials in order to enhance a particular property of concrete or by replacing the existing ingredient in order to reduce harmful effects of concrete.

The research work on concrete also includes utilization of some waste material as some percentage replacement of cement or aggregate. The objective of such type of work is to utilize waste material and thereby reducing the problems concerned with disposal of such waste and reducing the use of cement or aggregates. Replacement of cement by some other waste material has dual advantages from an environmental point of view: the reduction in amount of cement reduces the CO emissions and the utilization of waste material in place of cement addresses the issue of disposal of the waste, thereby reducing adverse effects on the environment. Reduction in cost is an added advantage of these types of concrete. If any material is added as ingredients of concrete, then investigations are to be carried out to check whether this added material doesn't influence the important properties of concrete in adverse manner. Hence, in general, the research work in the field of concrete needs an enormous amount of laboratory testing that involves cost, time, and human efforts.

Concrete is a widely used material, and various types of are being utilized and undergoing rapid evolution with recent developments in material science, manufacturing industries, and construction equipment and techniques, for fulfilling the need of being cost effective, ecofriendly, durable, and sustainable construction material. This clearly indicates the necessity of soft computing methods in the field of concrete. Application of artificial intelligence in the field of concrete is discussed in detail in the next section with respect to the two major areas:

- Prediction of mechanical properties of concrete, especially compressive strength
- Mix design of concrete

Data science and AI has been discussed in detail in Chapter 1 and 2; however, brief information about AI, especially the techniques that are applied in structural engineering are discussed here before presenting the application in prediction of properties of concrete and mix design.

4.4.2 ARTIFICIAL INTELLIGENCE (AI)

Artificial intelligence (AI) is widely recognized as a technology that provides an alternate method of solving challenging issues. AI has applications in various fields

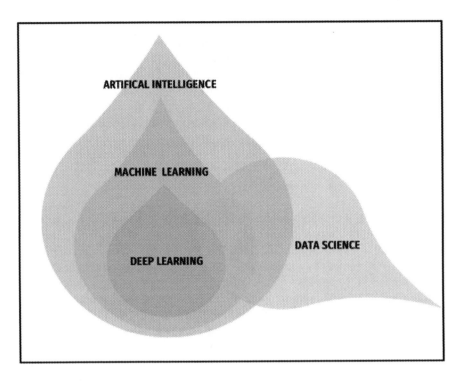

FIGURE 4.5 AI and Data Science.

such as engineering, medicine, military, marine, economy, and many more. AI mainly includes modeling and solves problems of identification, optimization, and prediction and control of complex systems.

Figure 4.5 shows the relation of important terms such as 'artificial intelligence', 'data science', 'machine learning', and 'deep learning'. The basics of these techniques are explained in Chapters 1 and 2 of this book.

ANN is a common artificial intelligence technique based on the analysis of biological neural networks that can be used in studies where a database of problems exists and the ANN model learns by doing. It has been shown to be an effective modelling technique for complex and nonlinear problems, with good learning and function approximation proposals. Artificial neural networks (ANNs) are modeled after the human brain and are biologically influenced. They are made up of a large number of basic processing elements known as neurons. ANNs are made up of clusters of neurons, which are also the fundamental unit of the system. An artificial neuron can be thought of as a basic signal converter. In several ways, the behavior of artificial neurons resembles that of neurons in the human brain. The following is a basic ANN example, which consists of three layers: the input layer, the secret layer, and the output layer are the three layers. Input variables are combined with neurons from the secret layer in the input layer. The output layer, on the other hand, contains the target data that the hidden layer must acquire. As a result, the entire learning process takes place in the secret layer, where relations between neurons are sought. A complex model can be built with a large number of neurons, which would be impossible to do with a simple design and so obscure that it would be difficult to construct a strictly empirical formula.

To determine the relationship between the inputs and outputs, the ANN needs some example patterns. It also needs to be able to generalize in order to predict unknown trends. All data should be partitioned into training, validating, and testing data sets during the partitioning process. Furthermore, the gathered data should be randomized to ensure the network's success. During the training process, a training algorithm is used to optimize the weights and biases of the ANN. It's possible to overlearn or overfit during this point. As a result, validating data is used to keep track of the training process and to halt network training if the validating error rises. The ANN model's forecasting ability is tested using the testing dataset after the training process is completed.

The data in the first fold is used for testing and validating phases in the first iteration of this process, while the data in the other folds is used to train the ANN. The second fold is used to evaluate and validate the ANN model in the second iteration. This process will be repeated until the kth round, and the average output of the obtained k ANN models will eventually reflect the ANN's efficiency.

4.4.3   APPLICATION OF AI FOR PREDICTION OF MECHANICAL PROPERTIES OF CONCRETE, ESPECIALLY COMPRESSIVE STRENGTH

A conventional method of determining mechanical properties of concrete is testing in laboratories or on-site. In the case of construction projects, regular testing of concrete is carried out, mainly tests to check workability of concrete and compressive strength of concrete by actually measuring it as per the standard procedures prescribed by the Indian Standard (IS) Codes. Apart from this testing, a huge amount of testing is required to be carried out while establishing various types of concretes with specific purposes.

Despite the fact that fresh and hardened concrete have a variety of specific properties, the workability and compressive strength of the concrete are the two most important characteristics of the concrete. The ease with which concrete can be handled, laid, and compacted is referred to as its workability. The 'slump cone test' is commonly used for measuring the workability of concrete. A few examples of applications of AI techniques for prediction of properties of concrete are discussed in the next section. These examples will give an idea, and how and which AI techniques are generally adopted.

4.4.3.1   Examples of Application of AI Techniques for Prediction of Concrete Properties

As discussed in section 4.2, in order to minimize the cost, time, and human efforts involved in experimentation, it is essential to work on soft computing models for prediction of various mechanical properties of various types of concrete. In general, it can be seen that artificial intelligent approaches have been largely used in the field of civil engineering since almost last two decades. Models have been developed using various AI techniques for the prediction of compressive strength of various types of concrete. The work has been carried out on various types of concrete and to find out various properties of it. Various soft computing methods such as genetic algorithms, regression analysis, fuzzy logic, and artificial neural networks to reduce experimental work and predict the desired performance parameter have been used. ANN is an important method for predicting and developing the input and output relation for complex problems, among the aforementioned approaches. Thus, it very evident that ANN has been used in a number of studies to estimate the compressive strength of concrete containing alternative materials in recent years.

In this section of the chapter, a few examples of the work carried out in this context are presented with sufficient details so as to understand the application of AI for prediction of mechanical properties of concrete.

Case Study 1
- **Problem Statement:** Predication of Compressive Strength of Geopolymer Concrete
- **Civil Engineering Aspect:**
  Geopolymer Concrete:
  In geopolyer concrete, waste fly ash from the thermal industry and wastewater from chemical refineries is used.

  Geopolymer concrete is created using a geopolymerization process that involves aluminosilicate components such as fly ash, metakaolin, and steel slag, as well as an alkaline liquid activator, such as sodium hydroxide and/ or sodium silicate, as well as an alkaline liquid activator. By using fly ash or steel slag in GPC manufacturing, CO_2 emissions were lowered by a significant amount when compared to cement.

  Experimental study has been used to determine the properties of geopolymer concrete by conventional methods; however, large-scale experimentation may not always be practical, and researchers may not always be able to investigate the relationships between input factors. As a result, artificial intelligence algorithms are used to forecast the compressive strength of geoploymer concrete.
- **AI Technique Techniques used:**
  1. Artificial neural network (ANN)
  2. Adaptive neuro fuzzy inference (ANFIS)
- **Methodology Adopted:**
  - Geoploymer concrete of grade M25, M30, and M35
  - Experimental work: 210 specimens tested for compressive strength
  - Input Parameters: fly ash, sodium, silicate solution, sodium hydroxide in solid state, and water contents.
  - Output Parameter: Compressive strength of geopolymer concrete
  - For ANN model, 'sigmoid function' was used to compute the weights of relationships between input and output parameters.
- **Conclusions:**
  - ANN and ANFIS are effective in prediction of compressive strength of geopolymer concrete.

- The comparative study of these two techniques shows that ANFIS is more effective.
- **Reference:** Dao et al. [52]

Case Study 2

- **Problem Statement:** Prediction of Compressive Strength of Recycled Aggregate Concrete (RAC)
- **Civil Engineering Aspect:**
  Recycled Aggregate Concrete (RAC)
- The waste generated during construction and demolition, particularly concrete waste, can be converted into recycled aggregates (RA), which can then be used in concrete mixes. Recycled aggregate concrete (RAC) is the name given to the concrete that has been manufactured by employing RA recycled aggregates (RA). Working on RA instead of aggregates as a natural resource helps to minimize the negative effects on the environment; hence, it is critical to continue working on it.
- **AI Technique Techniques used:**
  1. Multiple linear regression (MLR)
  2. Artificial neural network (ANN)
  3. Adaptive neuro fuzzy inference (ANFIS)
- **Methodology Adopted:**
  - Experimental work: 257 specimens tested for compressive strength
  - Input Parameters: 14. Cement, natural fine aggregates, recycled fine aggregates, natural coarse aggregates 10 mm, natural coarse aggregates 20 mm, recycled coarse aggregates 10 mm, recycled coarse aggregates 20 mm, admixture, water, water cement ratio, sand aggregate ratio, water to total material ration, aggregate cement ratio, replacement ratio of recycled aggregate fly ash, sodium, silicate solution, sodium hydroxide in solid state, and water contents.
  - Output Parameter: Compressive Strength of RAC
- **Conclusions:**
  - It was discovered that the ANN and ANFIS models were effective in forecasting the compressive strength of concrete after 28 days; however, the MLR model was found to be insufficiently effective for the same predicting purposes.
- **Reference:** Khademi et al. [53]

Case Study 3

- **Problem Statement:** Prediction of Modulus of Rupture of Concrete for Various Mixes
- **Civil Engineering Aspect:**
  Modulus of rupture is one of the important properties of every material. Modulus of rupture of concrete can be determined by carrying out flexural tests in the laboratory. As concrete is combination of various ingredients such as cement, sand, aggregate, and water; the properties of concrete mix depend on the quality as well as quantities of these ingredients. Hence

every time for determination of modulus of rupture of concrete, testing is needed, AI techniques proves to be effective here, thereby reducing time, cost, and manpower required for testing.

- **AI Technique Techniques Used:**
  1. Multiple linear regression (MLR)
  2. Artificial neural network (ANN)
  3. Adaptive neuro fuzzy inference (ANFIS)
- **Methodology Adopted:**
  - Experimental work:
    For training of network historical data was used. The laboratory testing in flexure of the $150 \times 150 \times 600$ mm prototype concrete beam specimens prepared from 14 different concrete mixtures. Two beams were prepared from each mix, tested, and their average values recorded as the results. For validation of model, the experimental data was used.
  - Input Parameters: Concrete mix proportions
  - Output Parameter: Modulus of rupture of concrete
  - Feed-forward back propagation learning algorithm is used, MATLAB software is used.
- **Conclusions:**
  - The researchers have reported that the ANN model was successful in predicting the modulus of rupture of concrete with minimum error less than 4%.
- **Reference:** David et al. [54]

Case Study 4
- **Problem Statement:** Prediction of Compressive Strength of High-Performance Concrete Containing Industrial By-products
- **Civil Engineering Aspect:**
  High-performance concrete:
  High-performance concrete is defined as concrete with strength and durability significantly beyond those obtained for regular concrete. The high performance is achieved with the use of low w/c (0.20–0.35) as well as pozzolans to produce a dense microstructure that is high in strength and low in permeability. A superplasticizer is added to keep the mix workable.

  With high cement content, the use of superpasticizers and silica fume and the need for more stringent quality control the unit cost of high-performance concrete can exceed that of normal concrete.
- **AI Technique Techniques used:**
  Artificial neural network (ANN)
  Model 1: with one hidden layer
  Model 2: with two hidden layer
- **Methodology Adopted:**
  - Experimental work:
  - Experimental results of 71 specimens with 14 different mixes tested by authors
  - Input Parameters: Concrete mix proportions

- Output Parameter: Modulus of rupture of concrete
- The statistical tools such as root mean square (RMS), mean square error (MSE), mean absolute percentage error (MAPE) and R2 were determined to check the capability of both the models for predication of compressive strength.
- **Conclusions:**
  - ANN models successfully predict the compressive strength of high-performance concrete containing industrial by-products.
- **Reference:** Vidivelli et al. [55]

The compressive strength of concrete is determined to be the subject of the majority of research in the prediction of mechanical properties of concrete. Compressive strength of concrete is acknowledged as one of its most significant mechanical properties. It is also recognized as one of the most important aspects in ensuring the quality of concrete. Pradeep Kumar and Sharma [56] on the other hand, worked on the slump value of concrete. The workability of the concrete mix can be determined by performing a slump test. It is defined as a property of concrete which determines the amount of effort requires to place, compact, and finish a concrete member. The slump cone test on concrete is described in detail in IS 456:2000. Pradeep Kumar and Sharma [56] used data from 500 tests performed at a ready mix concrete (RMC) factory in India to construct an artificial neural network (ANN) model for predicting slump value. After careful consideration, it has been determined that the model can provide optimal performance or can predict any mix proportions (resulting in the appropriate or desired slump), provided that their type of cement, their type of aggregates, and their types of constituents are within the range of data used for training [56].

All of the work presented above demonstrates that the application of artificial intelligence approaches for the prediction of mechanical properties of concrete, particularly compressive strength, is effective and that these techniques can be utilized to advance the field of concrete technology.

Focusing on the compressive strength of concrete, actual testing is carried out in order to evaluate it using the conventional approach. This results in concrete cubes with dimensions of $150 \times 150 \times 150$ mm being produced, and they are then allowed to cure for 28 days before being tested on a compressive testing machine (CTM) or a universal testing machine (UTM) (UTM). Using experimental data, soft computing models are created using a variety of approaches, and the models are then exposed to validation, as is the case with hard computing. Once these models have been created, they can be utilized to calculate the desired attribute in further depth. The procedure is clearly illustrated in Figure 4.6, which is easy to understand.

If we check Figure 4.6 critically, it will be well understood that, the 'input' and 'output' is the civil engineer's domain and the 'model' can be the computer engineer's expertise; hence, the entire problem becomes interdisciplinary. However, the 'model' part can be handled by a civil engineer. The entire efficiency of the model depends upon the relevant and reliable data input. Hence, it is the role of the civil engineer to study and completely understand the data for input. The model is made more and more accurate by reducing error by various techniques. Again, the validity of 'output' is to be critically checked by the civil engineer.

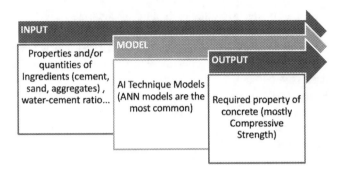

FIGURE 4.6　General Representation of Process.

Prediction of properties of concrete is one of the areas where AI techniques can be adopted successfully; another major area is mix design of concrete. In the next section, mix design and application of AI in mix design is discussed in detail.

4.4.4　APPLICATION OF AI TECHNIQUES IN MIX DESIGN OF CONCRETE

The mix design of the concrete is an essential and critical part of concrete technology. As concrete strength and performance depends on the quality of its ingredients, it is equally affected by the quantities of ingredients as well. In this subsection of the chapter, we will discuss in detail the significance of mix design, conventional methods of mix design, and application of artificial intelligence approaches in mix design in greater detail.

4.4.4.1　Mix Design of Concrete

A good and sound infrastructure is built on the foundation of a well-designed concrete mix. The quality of concrete in both its fresh and hardened states is determined not only by the workmanship, but also by the design of the concrete mix used in its production. Concrete mix design is the process of selecting suitable materials for concrete and determining their most optimal proportions in order to generate an economical concrete that meets the specified compressive strength, workability, and durability requirements while being cost-effective. Concrete with specified strength is produced by mixing the proper proportions of its constituents, which include cement, fine aggregates, coarse aggregates, water, and admixtures, among others. IS 456:2000 [51] specifies numerous types of concrete based on the grade of concrete used in the construction (grades of concrete). Specifically, according to IS 456:2000 [51], ordinary concrete is defined as concrete with an M10 to M20 grade, standard concrete is defined as concrete with an M25 to M55 grade, and high strength concrete is defined as concrete with a grade of M60 to M80. Conforming to the International Standard for Concrete Mixes (IS456:2000) [51], M refers to a mix and a number relates to the stipulated compressive strength of a 150 mm size cube after 28 days stated in N/mm^2. The compressive strength of concrete is determined by the proportions of the various elements in the concrete. Mix design is in charge of determining the proportions. Design mix is preferable above nominal mix, according to the International Standard 456:2000 [51]. It further stipulates that the function object [native code], in his or her capacity as guarantor of the quality of the concrete used in the construction, is responsible for the mix design. As a result, one of the most important areas in the world of concrete is the design of concrete mixes.

4.4.4.2 Methods of Mix Design of Concrete

The main objective of mix design is to satisfy the requirements of fresh concrete (workability) and properties of hardened concrete (strength and durability). Different design methods are followed all over the world for deciding the appropriate mix proportion. Some of the common mix design methods are:

1. Mix design according to Indian Standard Recommended guidelines
2. ACI mix design method
3. High strength mix method
4. Road Note no 4 mix design method.

All of these methods work on the sets of tables, graphs for calculating density, water requirements, and aggregate and cement requirements. These methods are discussed here in brief.

1. Mix Design According to Indian Standard:
   IS 10262:2009 [57] specifies the standards for calculating the proportions of concrete mixes to meet specific specifications by utilizing concrete-making components. Bureau of Indian Standards has proposed the Indian Standard IS 10262:2009 [58], which is an international standard. In order to accomplish specific properties such as strength at a particular age, workability of new concrete and durability requirements, the proportioning process is carried out [58].

2. ACI Mix Design Method:
   An effective technique of concrete mix design, taking into account the more affordable use of locally available ingredients to generate acceptable workability, durability, and strength, has been advocated by the American Concrete Institute (ACI). The ACI method is capable of producing concretes with varying degrees of workability, ranging from extremely stiff to fluid state workability, depending on the conditions.

   Among the many advantages of the ACI method, which is widely used in the United States, is its simplicity, which is exemplified by the fact that it applies equally well and with a largely identical procedure to round or angular aggregate, to normal or lightweight aggregate, and to air-entrained or non-air-entrained concrete. Based on the fact that, for a given size of well-graded aggregates, water content is essentially independent of mix proportions, that is, water content remains constant regardless of variation in water cement ratio and cement content, the ACI technique is used. However, regardless of the shape of the particles, this method assumes that even after complete compaction is completed, a certain percentage of air

remains that is inversely proportional to the maximum size of aggregate. This method also assumes that even after complete compaction is completed, a certain percentage of air remains that is inversely proportional to the maximum size of aggregate [58].

3. High Strength Mix Method:

   Shacklock and Erntroy developed the high-strength concrete mix design in 1954, which was first used in the United Kingdom. The strength of fully compacted concrete at a required age is believed to be solely reliant on the water-to-cement ratio (w/c) of the concrete mix when building concrete mixes with low and medium grade compressive strengths, i.e., up to 35 MPa. However, the compressive strength of high-strength mixes greater than 35 MPa is mostly affected by the characteristics of the aggregates, rather than the weight-to-compaction ratio. As a result, the methods of mix design used for medium grade concrete cannot be relied on to provide an accurate estimate of the needed mix proportions for high strength concrete under all conditions, and consequently cannot be utilized to design high strength concrete [58].

4. Road Note no 4 Mix Design Method:

   The aggregate to cement ratios are calculated using this approach based on the type of aggregate used, the maximum size of the aggregate used, and the varying levels of workability of the aggregate. The combined grading curves are used to calculate the relative amount of aggregates in a mixture. The usage of multiple types of fine and coarse aggregates in the same mix is made easier using this technique. The proportional proportion of each of these can be estimated with reasonable ease using a combination of grading curves. The values of the aggregate to cement ratio are provided for coarse aggregate that is angular, rounded, or irregular in shape. One of the disadvantages of this process is that it results in extremely high cement content, which is why it is becoming outdated. Using gap graded aggregate becomes unavoidable in many situations. Most of the country follows the practice of using coarse aggregates no larger than 20 mm rather than smaller than 10 mm. As a result of the extremely low quality of 10 mm aggregates produced by the jaw crusher, this occurs. It is not possible to incorporate gap grading into the usual combined grading curves of the RRL approach. There are several grades of sand accessible in different sections of the country, but the coarse fraction (1.18 mm and above) and the fine fraction (below 1.18 mm) are the most common (600 micron and below). It is difficult to alter the sand content of a concrete mix to match any of the conventional combined grading curves that are available. It is common in such situations for the combined grading curve to cut across more than one standard curve. Different aggregate to cement ratios are offered for different levels of workability ranging from low to high. However, unlike previous approaches, these levels of workability are not specified in terms of slump, compaction factor, or Vee be time. Because the fine aggregate content cannot be altered to account for variations in cement content, the sand proportions in both the richer and leaner mixes may be the same for a given set of components [58].

4.4.4.3 Examples of Application of AI Techniques for Mix Design of Concrete

As mentioned in Section 4.4.4.2, the main objective of mix designs is to maintain the workability as well as the strength and durability of a mix. Hence, to achieve this, the advanced techniques must be inculcated while designing the concrete mix. It is one of the major challenges to maintain the workability and the strength of the mix which can be overcome by means of AI technique.

Case Study 1
- **Problem Statement:**
  - To train and evaluate an ANN model for concrete mix design using standard compressive strength of concrete in order to improve the design of concrete mixes.
- **Civil Engineering Aspect:**
  Concrete is made up of four fundamental components: cement, fine aggregate, coarse aggregate, and water. Cement is the most common type of material used in concrete production. All of the factors that go into the design of a concrete mix, including the grade of the concrete, the cement used, and the size, shape, and grading of aggregates, are important considerations. It is possible to create concrete mixes using an empirical relationship between the components that influence the choice of mix design, which has been discovered through experimentation in the field of concrete mix design. Primary disadvantages of this method include that it does not produce the desired strength, the calculations are time-consuming and require consultation of several tables in order to arrive at the trial mix proportion, and the variation in attainment of desired strength is uncertain below target strength and may even fail completely. Additional disadvantages include that it is expensive and time-consuming to implement (Rama Shanker 2014). It is possible to combat this problem by including ANN into the process of constructing concrete mixtures.
- **AI Technique Technique Used:**
  1. Artificial neural network (ANN)
- **Methodology Adopted:**
  - Back propagation ANN trained neural network is integrated in the model to learn experimental data to predict 7, 14, 28 days of compressive strength of concrete, which have been loaded in model, containing 207 concrete mixtures.
  - Ingredients proportioning of concrete to achieve the desired strength at site.
  - Comparison of the results of conventional method of concrete mix to the software analysis: comparison on the basis of results available on both from software as well as site is done.
- **Conclusions:**
  - ANN model is proposed for proportioning of concrete mix design.
  - Application of artificial intelligence in the field of mix design is very appropriate in order to preserve valuable time at reasonable cost.

- Reduction in the number of trials in laboratory as well as field, cost savings of material as well as labor and also time saving is achieved using ANN as it provides higher accuracy.
- Higher durability of the concrete designed is expected and hence found economical.
- **Reference: Shah et al.** [59]

Case Study 2
- **Problem Statement:** To propose the artificial neural network (ANN) model for approximate proportioning of concrete mixes.
- **Civil Engineering Aspect:**
  Concrete is characterized by the type of aggregate or cement used, by the specific qualities it manifests, or by the methods used to produce it. In ordinary structural concrete, the character of the concrete is largely determined by a water-to-cement ratio. The lower the water content, all else being equal, the stronger the concrete. The mixture must have just enough water to ensure that each aggregate particle is completely surrounded by the cement paste, that the spaces between the aggregate are filled, and that the concrete is liquid enough to be poured and spread effectively. It becomes quite difficult to balance the workability and the strength of concrete and so AI is used.
- **AI Technique Technique Used:**
  1. Artificial neural network (ANN)
- **Methodology Adopted:**
  - Database of about 55 mixes retrieved from various literatures to predict the results from ANN models.
  - Experimental work: 55 specimens tested for compressive strength for 7, 14, and 28 days based on five input parameters such as cement, sand, coarse aggregate, and water and fineness modulus.
  - Input Parameters: Cement, sand, aggregates, and water
  - Output Parameter: Artificial neural network (ANN) model for approximate proportioning of concrete mixes.
  - Back propagation method of training of ANN. It is a supervised learning method and is a generalization of the delta rule.
- **Conclusions:**
  - The ANN model is proposed for proportioning of concrete mix design.
  - The model reduced the number of trials in laboratory as well as field, saved cost of material as well as labor, and also saved time as it provides higher accuracy.
  - The concrete designed is expected to have higher durability and hence is found economical.
- **Reference: Gupta** [60]

Case Study 3
- **Problem Statement:** To develop a two-step computer-aided approach for pozzolanic concrete mix designs.

- **Civil Engineering Aspect:**
  Pozzolanic Concrete
- Pozzolans are silicate-based materials that react with (consume) the calcium hydroxide produced by hydrating cement to form more cementitious materials. Pozzolans are used in the production of a variety of cementitious materials. Pozzolans mix with lime to form extra calcium silicate hydrate, which is the substance responsible for binding concrete together during construction. They are utilized in the production of masonry mortars and plastering. Because it produces a better surface polish, it is employed in the construction of decorative and art structures, as well as the production of precast sewage pipes.
- The addition of natural pozzolanic materials to concrete reduced the amount of "hazardous components" present in the concrete as a result of the natural pozzolan's reaction with them, which varied depending on the fineness of the natural pozzolan (Ükrü Yetgin1 and Ahmet avdar 2006). Natural pozzolans, in addition to their capacity to quickly divide, can fill the micropores in the cement matrix and considerably increase the durability of cements by altering the structural framework of the matrix (Sabir et al. 2001; Shannag 2000; Pan et al. 2003). Because of their tiny grain size, natural pozzolans also have lubricating effects on cement mortar or concrete; they improve the consistency of the concrete.
- **AI Technique Technique Used:**
  1. Artificial neural network (ANN)
- **Methodology Adopted:**
  - Develop a data set of pozzolanic concrete mixture proportioning that is compliant with the American Concrete Institute code, which includes experimental data collected from the literature as well as numerical data generated by a computer program in which artificial neural networks (ANNs) are used to develop prediction models for compressive strength and slump of the concrete.
  - Twelve distinct combinations of experimental specimens were used to test the ANN models, which were tested using the data from the specimens. Classification of the dataset pertaining to pozzolanic concrete mixture proportioning methods. A classification method is used to divide the dataset into 360 classes based on compressive strength, pozzolanic admixture replacement rate, and material cost. The data set is divided into 360 classes using the classification method.
- **Conclusions:**
  - Developed a two-step computer-aided approach for pozzolanic concrete mix design.
  - The proposed computer-aided approach is found out to be convenient for pozzolanic concrete mix design and practical for engineering applications.
- **Reference: Kao et al.** [61]

Case Study 4
- **Problem Statement:** To use artificial neural networks (ANNs) for the checking of design composition of fly ash blended cement concrete mixes and to check its compressive strength.
- **Civil Engineering Aspect:**
  Fly Ash Blended Cement Concrete
  Mixing fly ash with Portland cement clinker can be done in two ways: either by intergrinding the fly ash with the clinker or by mixing dry fly ash with the portland cement clinker. When used in Portland cement concrete (PCC), fly ash provides several benefits as well as an improvement in concrete's performance in both the fresh and hardened states. It has been shown that the inclusion of fly ash in concrete increases the workability of plastic concrete as well as the strength and durability of hardened concrete. The usage of fly ash is also a cost-effective option. When fly ash is added to concrete, it is possible to reduce the amount of portland cement used. When measured at equilibrium conditions in different relative humidities, the porosity of fly ash blended cement mortar specimens was found to be 10% to 15% lower when compared to the porosity of OPC mortar specimens. As a result, the compressive strength of fly ash blended cement mortar specimens was found to be 6% to 8% higher than that of OPC mortar specimens.
  AI Technique Technique Used:
  1. Artificial neural network (ANN)
- **Methodology Adopted:**
  - Use of a simple back propagation neural network to model the design mix proportions for the fly ash blended concrete mixes.
  - Analyses of data in the network fitting tool of MATLAB computer software.
  - Regression analysis used to assess the performance of fitting tool which is required to map numeric inputs into targets.
- **Conclusions:**
  - The compressive strength values of fly ash blended concrete mixes can be predicted with the help of artificial neural network concepts.
  - ANN may be used for verification of a theoretical concrete mix design procedure specified in standard guidelines.
  - Decisions at appropriate times can be taken, cost can be reduced by reducing wastages and speed of construction can be improved with the help of prediction of compressive strength by ANN concepts.
- **Reference: Alok Verma and Ishita Verma** [62]

Case Study 5
- **Problem Statement:**
  - To utilize state-of-the-art achievements in machine learning techniques for concrete mix design.
  - To estimate the compressive strength of the concrete mix resulting from a specific composition of concrete mix ingredients.

- To transform ANN into an actual mathematical equation, which can be used in engineering practice.
- **Civil Engineering Aspect:**
  Concrete Mix
  Concrete holds a unique position among the current construction materials in terms of durability. Concrete is a building material that is composed of a hard, chemically inert particle element known as an aggregate (which is often composed of several types of sand and gravel) that is joined together by cement and water to form a structurally sound structure. Concrete mix design is the process of selecting suitable materials for concrete and determining their relative proportions with the goal of generating a concrete with the required strength, durability, and workability at the lowest possible cost. It is also known as concrete formulation design (Jay H Shah, Sachin B Shah 2014).
- **AI Technique Technique Used:**
  - Machine learning – Artificial neural network (ANN)
- **Methodology Adopted:**
  - Feeding of extensive database that consist of 741 records of concrete recipes to estimate the compressive strength.
  - Division of database into three subsets as per the specificity of machine learning which were split up as follows: The training materials 2019, 12, 1256 13 of 16 subset has 395 records (53.3%), the selection subset has 133 records (17.9%), and the testing subset has 134 records (18.1%).
  - Calculation of the suitable training rate and the step for the quasi-Newton training direction by the Broyden–Fletcher–Goldfarb–Shanno algorithm and the Brent method, respectively.
  - Regression analysis is used to assess the performance of fitting tool which is required to map numeric inputs into targets.
  - Transformation of the ANN into an actual mathematical equation, which can be used in engineering practice.
- **Conclusions:**
  - The compressive strength of the concrete mix resulting from a specific composition of concrete mix ingredients is estimated.
  - ANN transformed into an actual mathematical equation, which can be used in engineering practice.
- **Reference: Patryk Ziolkowski; Maciej Niedostatkiewicz [63]**

4.5 CONCLUSION AND FUTURE SCOPE

Civil engineering and its allied fields are developing rapidly due to the ever-increasing demand for the construction sector. Cities, metropolitan landscapes, and heavily industrialized zones that are rapidly expanding are stimulating the rise of the infrastructure industry, which in turn is fueling the growth of supporting infrastructure such as bridges, highways, and networking services, among other types of infrastructures. This expansion is clearly visible in all countries, and it has been demonstrated to be increasing momentum on a daily basis. When it comes to the development of these sectors, civil engineers rely heavily on structural engineering components, and it is a rising domain within the field of civil engineering itself. Modern building methods and materials not only necessitate excellent design, but they should also be concerned with cost optimization, which may be accomplished through a variety of means. The use of appropriate tools and approaches during the design process will help to reduce waste of materials while increasing the overall sustainability. With the introduction of computing technology and the ever-growing number of computing tools available, the construction sector now has actual help in the form of appropriate approaches such as big data–based processing, artificial intelligence–based design, and so on. The application of artificial intelligence approaches to forecast the strength of materials is now being researched and tested in the field for product and usage. AI is a vast field that encompasses a plethora of approaches, tools, processes, and protocols; picking the most appropriate method from among them is the key to efficiently utilizing AI technology. When it comes to the tools accessible for use, ANN is one of the most common ways that is being used widely by a plethora of researchers all over the world. It is easy to use and may be used for a variety of applications, including prediction, segregation and classification, among others. Because it is a versatile tool, it not only assists designers in conveying complicated material properties in a straight-forward manner, but it also contributes to the development of sustainable infrastructure solutions. As a result, the use of ANN and other comparable tools for material creation is highly recommended; there is tremendous potential for its application given the continually growing nature of the construction sector.

REFERENCES

1. IS 875 Part 2 -1987 Reaffirmed 2003, Code of Practice for Design Loads (Other than Earthquake) for Buildings and Structures.
2. "Mumbai CST bridge collapse LIVE updates: Six dead, 30 injured; CM orders inquiry". *The Indian Express*. 15 March 2019.
3. "Watch: Part of Rs 263 Crore Bihar Bridge Collapses", *Outlook, Saturday*, Oct 09, 2021.
4. "2 Killed, 7 Injured after Under-construction Bridge Collapses in Bengal's Malda District", *News18*, First Published: February 16, 2020.
5. Goldberg, D. E. "*Genetic Algorithms in Search, Optimization and Machine Learning*". Longman Press, USA (1989).
6. Holland, J. H. "*Adaptation in Natural and Artificial Systems*". University of Michigan Press, USA (1975).
7. Jong, K. A., Dejong, K. "*The Analysis of the Behavior of a Class of Genetic Adaptive Systems*". The University of Michigan Press, USA (1975).
8. Rich, E., Knight, K. "*Artificial Intelligence*". McGraw-Hill, USA (1996).
9. Silva, M., Santos, A., Figueiredo, E., Santos, R., Sales, C. et al. "*A Novel Unsupervised Approach Based on a Genetic Algorithm for Structural Damage Detection in Bridges*". Engineering Applications of Artificial Intelligence (2016).
10. Mroginski, J. L., Beneyto, P. A., Gutierrez, G. J., Di, R. A. "*A selective genetic algorithm for multiobjective optimization of cross sections in 3D trussed structures based on a spatial sensitivity analysis*". *Multidiscipline Modeling in Materials and Structures*, vol. 12, no. 2, (2016), pp. 423–435.

11. Truong, V. H., Nguyen, P. C., Kim, S. E. *"An efficient method for optimizing space steel frames with semi-rigid joints using practical advanced analysis and the micro-genetic algorithm"*. Journal of Construction Steel Research, vol. 128, (2017), pp. 416–427.

12. Artar, M., Daloglu, A. T. *"Optimum weight design of steel space frames with semi-rigid connections using harmony search and genetic algorithms"*. Neural Computing & Applications, vol. 29, no. 11, (2018), pp. 1089–1100.

13. Zarbaf, S. E. H. A. M., Norouzi, M., Allemang, R. J., Hunt, V. J., Helmicki, A. *"Stay cable tension estimation of cable-stayed bridges using genetic algorithm and particle swarm optimization"*. Journal of Bridge Engineering, vol. 22, no. 10, (2017), pp. 1–8.

14. Mehrkian, B., Bahar, A., Chaibakhsh, A. *"Semiactive conceptual fuzzy control of magnetorheological dampers in an irregular base-isolated benchmark building optimized by multi- objective genetic algorithm"*. Structural Control & Health Monitoring, vol. 26, (2019), pp. 1–28.

15. Hackwood, S., Beni, G. *"Self-organization of sensors for swarm intelligence"*. IEEE International Conference on Robotics and Automation, vol. 1, (1922), pp. 819–829.

16. Bonabeau, E., Dorigo, M., Theraulaz, G. *"Swarm Intelligence: From Natural to Artificial Systems"*. Oxford University Press, USA (1999).

17. Saleem, M., DiCaro, G. A., Farooq, M. *"Swarm intelligence based routing protocol for wireless sensor networks: survey and future directions"*. Information Sciences, vol. 181, no. 20, (2011), pp. 4597–4624.

18. Lu, P. Z., Chen, S. Y., Zheng, Y. J. *"Artificial intelligence in civil engineering. Mathematical Problems in Engineering"* (2012).

19. Kennedy, J., Eberhart, R. *"Particle swarm optimization"*. Proceedings of the 1995 IEEE International Conference on Neural Networks, vol. 4, (1995), pp. 1942–1948.

20. Xue, X. H., Xiao, M. *"Deformation evaluation on surrounding rocks of underground caverns based on PSO-LSSVM"*. Tunnelling and Underground Space Technology, vol. 69, (2017), pp. 171–181.

21. Chen, X. L., Fu, J. P., Yao, J. L., Gan J. F. *"Prediction of shear strength for squat RC walls using a hybrid ANN-PSO model"*. Engineering with Computers, vol. 34, no. 2, (2018), pp. 367–383.

22. Qin, S. Q., Zhang, Y. Z., Zhou, Y. L., Kang, J. T. *"Dynamic model updating for bridge structures using the kriging model and PSO algorithm ensemble with higher vibration modes"*. Sensors, vol. 18, no. 6, (2018), pp. 1879.

23. Karaboga, D. *"An Idea Based on Honey Bee Swarm for Numerical Optimization"*. Erciyes University, Turkey (2005).

24. Kang, F., Li, J. J. *"Artificial bee colony algorithm optimized support vector regression for system reliability analysis of slopes"*. Journal of Computing in Civil Engineering, vol. 30, no. 3, (2016) 04015040.

25. Yang, X. S. *"Nature-Inspired Metaheuristic Algorithms"*. Luniver Press, UK (2008).

26. Gandomi, A. H., Yang, X. S., Talatahari, S., Alavi, A. H. *"Firefly algorithm with chaos"*. Communications in Nonlinear Science and Numerical Simulation, vol. 18, no. 1, (2013), pp. 89–98.

27. Lieu, Q. X., Do, D. T. T., Lee, J. *"An adaptive hybrid evolutionary firefly algorithm for shape and size optimization of truss structures with frequency constraints"*. Computers & Structures, vol. 195, (2018), pp. 99–112.

28. Kubair, K. S., Mohan, S. C. *"Numerical model updating technique for structures using firefly algorithm"*. International Conference on Recent Advanced in Materials, Mechanical and Civil Engineering, vol. 330, (2018), p. UNSP012123.

29. Yang, X. S., Deb, S. *"Cuckoo search via Lévy flights"*. Proceedings of World Congress on Nature and Biologically Inspired Computing, 2009, pp. 210–214.

30. Yang, X. S., Deb, S. *"Engineering optimization by cuckoo search"*. *International Journal of Mathematical and Numerical Optimisation*, vol. 4, (2010), pp. 330–343.
31. Yang, X. S., Deb, S. *"Multiobjective cuckoo search for design optimization"*. *Computers & Operations Research*, vol. 40, no. 6, (2013), pp. 1616–1624.
32. Zamani, A. A., Tavakoli, S., Etedali, S. *"Fractional order PID control design for semi-active control of smart base isolated structures: a multi-objective cuckoo search approach"*. *ISA Transactions*, vol. 67, (2017), pp. 222–232.
33. Rakhshani, H., Rahati, A. *"Intelligent multiple search strategy cuckoo algorithm for numerical and engineering optimization problems"*. *Arabian Journal for Science and Engineering*, vol. 42, no. 2, (2017), pp. 567–593.
34. Chalkoff, R. J. *"Artificial Neural Networks"*. McGraw-Hill Press, USA (1997).
35. Basheer, I. A., Hajmeer, M. *"Artificial neural networks: fundamentals, computing, design, and application"*. *Journal of Microbiological Methods*, vol. 43, no. 1, (2000), pp. 3–31.
36. McCulloch, W., Pitts, W. *"A logical calculus of the ideas immanent in nervous activity"*. *Bulletin of Mathematical Biophysics*, vol. 52, no. 1-2, (1943), pp. 99–115.
37. Adeli, H., Yeh, C. *"Preceptron learning in engineering design"*. *Microcomputer in Civil Engineering*, vol. 4, (1989), pp. 247–256.
38. Hecht-Nielsen, R. *"Neurocomputing"*. Addison-Wesley Press, UK (1990).
39. Hasanipanah, M., Noorian-Bidgoli, M., Armaghani D. J., Khamesi H. *"Feasibility of PSO-ANN model for predicting surface settlement caused by tunnelling"*. *Engineering with Computers*, vol. 32, no. 4, (2016), pp. 705–715.
40. Garcia-Segura, T., Yepes, V., Frangopol, D. M. *"Multi-objective design of post-tensioned concrete road bridges using artificial neural networks"*. *Structural and Multidisciplinary Optimization*, vol. 56, no. 1, (2017), pp. 139–150.
41. Cascardi, A., Micelli, F., Aiello, M. A. *"An artificial neural networks model for the prediction of the compressive strength of FRP-confined concrete circular columns"*. *Engineering Structures*, vol. 140, (2017), pp. 199–208.
42. Yan, F., Lin, Z. B., Azarmi, F., Sobolev, K. *"Evaluation and prediction of bond strength of GFRP-bar reinforced concrete using artificial neural network optimized with genetic algorithm"*. *Composite Structures*, vol. 161, (2017), pp. 441–452.
43. Chatterjee, S., Sarkar, S., Hore, S., Dey, N., Ashour, A. S. et al. *"Particle swarm optimization trained neural network for structural failure prediction of multistoried RC buildings"*. *Neural Computing & Applications*, vol. 28, (2017), pp. 2005–2016.
44. Naderpour, H., Rafiean, A. H., Fakaharian, P. *"Compressive strength prediction of environmentally friendly concrete using artificial neural networks"*. *Journal of Building Engineering*, vol. 16, (2018), pp. 213–219.
45. Kalogirou, S. A. *"Artificial neural networks in renewable energy systems applications: a review"*. *Renewable & Sustainable Energy Reviews*, vol. 5, no. 4, (2001), pp. 373–401.
46. Adeli, H. *"Neural networks in civil engineering: 1989-2000"*. *Computer-Aided Civil and Infrastructure Engineering*, vol. 16, no. 2, (2001), pp. 126–142.
47. Walker, S. *"Big data: a revolution that will transform how we live, work, and think"*. *International Journal of Advertising*, vol. 33, no. 1, (2014), pp. 181–183.
48. Tran, C. *"Structural-damage detection with big data using parallel computing based on MPSOC"*. *International Journal of Machine Learning and Cybernetics*, vol. 7, no. 6, (2016), pp. 1213–1223.
49. Wu, X. D., Zhu, X. Q., Wu, G. W., Ding, W. *"Data mining with big data"*. *IEEE Transactions on Knowledge and Data Engineering*, vol. 26, no. 1, (2014), pp. 97–107.
50. *"A concrete contribution to the environment ™"*, *EcoSmart Concrete™*, http://ecosmartconcrete.com/

51. IS 456: 2000 Plain and Reinforced Concrete -Code of Practice.
52. Van Dao, D., Ly, H.-B., Trinh, S. H., Le, T.-T., Pham, B. T. *"Artificial intelligence approaches for prediction of compressive strength of geopolymer concrete"*. *Materials* vol. 12, (2019), p. 983. doi: 10.3390/ma12060983, (2019).
53. Khademi, F., Jamal, S. M., Deshpande, N., Londhe, S. *"Predicting strength of re-cycled aggregate concrete using artificial neural network, adaptive neuro-fuzzy inference system and multiple linear regression"*, *International Journal of Sustainable Built Environment*, vol. 5, (2016), pp. 355–369. 10.1016/j.ijsbe.201 6.09.003.
54. David, O. O., Chioma, A. T. G. *"Artificial neural network for the modulus of rupture of concrete"*, *Advances in Applied Science Research*, vol. 4, no. 4, (2013), pp. 214–223.
55. Vidivelli, Dr. B., Jayaranjini, A. *"Prediction of Compressive Strength of High-Performance Concrete Containing Industrial by Products Using Artificial Neural Networks"*, *International Journal of Civil Engineering and Technology (IJCIET)* vol. 7, no. 2, (2016), pp. 302–314.
56. Kumar, P., Sharma, Dr. I. C. *"Modelling Ready Mix Concrete Slump using Artificial Neural Network"*. *IJLTEMAS*, vol. III, no. VII, (2014), ISSN 2278–2540, pp. 284–288.
57. IS 10262:2009 Concrete Mix Proportioning – Guidelines, Bureau of Indian Standards.
58. Shetty, M. S. *"Concrete Technology"*, *S. Chand* (2001).
59. Shah, S. V., Pawar, Ms D. A., Patil, Ms A. S., Bhosale, Mr P. S., Subhedar, Mr A. S., Bhosale, Mr G. D. *"Concrete mix design using artificial neural network"*, *International Journal of Advance Research in Science and Engineering*, vol. 07, Special Issue No 03, (2018), pp. 251–259.
60. Gupta, S. *"Concrete mix design using artificial neural network"*, *Journal on Today's Ideas –Tomorrow's Technologies (JOTITT)*, vol. 01, no. 01, (2013). DOI: 10.15415/ jotitt.2013.11003.
61. Kao, C.-Y., Shen, C.-H., Jan, J.-C., Hung, S.-L. *"A computer-aided approach to pozzolanic concrete mix design"*, *Hindawi Advances in Civil Engineering*, vol. 2018, (2018), Article ID 4398017, 10.1155/2018/4398017.
62. Verma, A., Verma, I. *"Use of artificial neural network in design of fly ash blended cement concrete mixes"*, *International Journal of Recent Technology and Engineering*, vol. 8, no. 3, (2019), pp. 4222–4233.
63. Ziolkowski, P., Niedostatkiewicz, M. *"Machine learning techniques in concrete mix design"*, *MDPI Materials*, vol. 12, (2019), p. 1256. doi: 10.3390/ma12081256.

5 Application of Data Science in Transportation Systems

Rakesh K. Jain, Ashish R. Joshi, and Bharati H. Gavhane

5.1 INTRODUCTION TO TRANSPORTATION ENGINEERING

Transportation engineering is a subset of civil engineering engaged in any nation's industrial, economical, cultural, and social development. It combines scientific fundamental and technological advancement to plan, functionally design, build, operate, and manage various transportation modes to offer the secure, proficient, rapid, accessible, and cost-effective movement of human beings and freight. The importance of transportation for a nation is comparable to the veins inside the human body. As the veins facilitate the health of humans, on the same line, transportation preserves the health of a nation by keeping the humans and other resources moving from one region to some other vicinity. The transportation as a system is diverse and multifaceted because it handles complexities related to multi-modal, multi-sectoral, multi-problem, multi-purpose, and multi-discipline, as depicted in Figure 5.1.

5.1.1 SUBDIVISION OF TRANSPORTATION ENGINEERING

The civil engineer as a transportation expert usually entails planning, design, building, operation, and maintenance of transportation infrastructures and facilities for various modes such as air, road, railroad, pipeline, water, space transportation, and even telecommunication.

Transportation engineering involves forecasting, planning, decision making, and developing various facilities considering socioeconomic and political factors. The forecasting of passenger travel entails transport modeling, requiring the assessment of trip generation (number and purpose of trips), trip distribution (origin and destination study), mode choice (transport mode used), and assignment of the route (route selected for the trip). The forecasting can also consider factors such as vehicle ownership, linking of the trips, and location choice of land use for various purposes such as residential or commercial. Designing transportation means deciding the size of infrastructure and facilities such as capacity and number of lanes, determining the type of material, its thickness used in pavement, and geometric design (alignment of vertical and horizontal curves) of the road or railway.

DOI: 10.1201/9781003316671-5

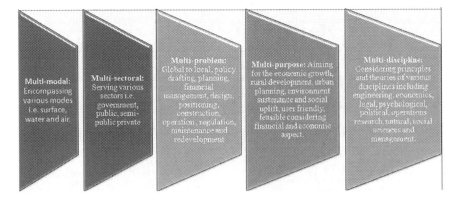

FIGURE 5.1 Diverseand Multifaceted Transportation System.

Along with these functions, plan preparation, analysis, finance, logistics, and policy evaluation are also crucial to transport engineers. In this domain, engineering principles are applied to enhance the transportation system by utilizing the three design controls: the drivers, the vehicle, and the mode (such as roadway, railway, etc.) itself (L.R. Kadiyali, 2007).

5.1.1.1   Highway Engineering

Highway engineering involves planning, designing, building, and operating highway roads and systems, urban streets, and parking centers. Highway engineer plays an important role in finalizing the most suitable route from the available alternatives considering various aspects such as finance, environmental impact evaluation, and valuation by carrying out the comparative study. The highway engineering field is very wide and is concerned with diverse activities, as mentioned below:

- Highway planning and alignment of road systems in the plain country and hilly areas based on location, traffic, and function. This system includes expressways, national highways (freeways), state highways, district roads, arterial roads, collector streets, and rural roads.
- The geometric design of the highway is concerned with the physical proportion and layout of visible features of the road.
- Traffic engineering involves the study of traffic volume and its characteristics for smooth flow, regulation, controlling, guiding, administration, and management to have a safe, convenient, rapid, and economical transport and goods.
- Pavement and roadway engineering provide satisfactory surface by the suitable structural design of road layers for operation of vehicular traffic.
- Intelligent transportation system.
- Design and construction of bridges, retaining structures, subways, and tunnels.

5.1.1.2 Railway Engineering

The railway is the fourth-largest transport network globally. It is a subset of transportation engineering involved with the planning, designing, construction, operation, and maintenance of railways considering various aspects of civil, mechanical, communication, and electrical engineering. This specialization includes:

- Setting horizontal and vertical alignment of the track.
- Finalizing station position and its layout.
- Estimation and costing of the project.
- Establishing signaling and controlling system.
- Focusing on the train movement.
- Building a cleaner and more secure transportation system utilizing, reinvesting, and revitalizing the rail network to cater to future needs.

5.1.1.3 Water Transportation: Port and Harbor

Water transportation is used to move people and cargo by ships, boats, barge, or ferry across ocean, sea, river, canal, or other modes. It is the most cost-effective and vital means for transporting across long distances. Port and harbor engineering is a major field of underwater transportation dealing with the planning, designing, building, and operating the ports, harbors, and other maritime centers. It also includes

- Meeting the need for water transportation, including its commercial and recreational aspects.
- Planning the future needs of water transportation projects, including environmental impacts of such development
- Erecting and handling various equipment and facilities on the port.

5.1.1.4 Airport Engineering

Airport engineering includes the planning, designing, and construction of airport facilities such as terminal building, runway, and various navigation systems for safe and efficient movement of passengers and goods. The role of airport engineer is:

- To finalize the orientation of the runway by analyzing predominant wind speed and direction.
- To fix the size of the runway and other features based on its function and future need.
- To ensure all the safety measures and exceptional clearances between wing-tips for all gates.
- To allocate the clear zones at the airport.

5.1.1.5 Pipeline Engineering

Pipeline engineering deals with the design, construction, and maintenance of pipes, stations for pumping, and facilities for storage. This mode is used to transport various fluids such as gas, petroleum products, and water for a long distance. In the water slurry form, pulverized coal and iron ore are also transported using this mode.

5.1.1.6 Telecommunications

The purpose of transportation and telecommunication is to connect to people without the constraint of physical distance. Transportation needs a mode such as a road or a rail to move public or freight from one place to another. Telecommunication involves e-mails, phones, messages, and letters for communication. Transportation and telecommunication both complement one another and may also be substituted for each other. Both contribute to economic growth, social connectivity, and the nation's development. Telecommunication is a high-capacity network with less constraint, including crossing various physical features and oceans by fiber-optic cables. Telecommunication uses an array of tools such as satellites, highly sophisticated computing networks, television, video and audio conferencing, phones, telegraph, radar, telemetry, and e-mail for transmitting big data and information throughout the globe. It is proved from data analytics that travel and communications both are growing simultaneously and probably stimulating each other.

The different modes of transportation mainly land transport, water transport, and air transport. Land transport includes pathways, roadways, and railways. Water transport consists of inland water transport and sea transport. Air transport includes

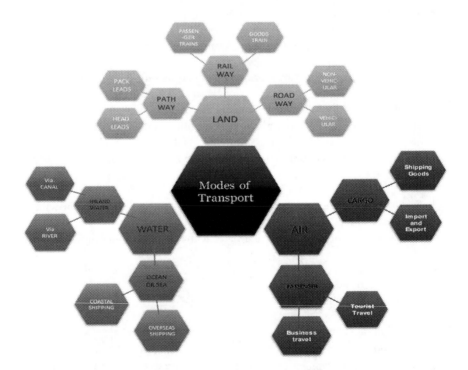

FIGURE 5.2 Modes of Transportation/Modal Characteristic (Modified after Filip Biljecki et al. 2013).

SCALE	ENVIRONMENTAL	HISTORICAL	TECHNOLOGICAL	POLITICAL	ECONOMICS
LOCAL	Hydrography and Geomorphology	Culture and Settlement Pattern	Roads	Zoning	Employment and Distribution
REGIONAL	Climate	Urban System	Railways and Cables	Taxations and Regulations	Modal Competition and Regulations
NATIONAL	Distance	National State Colonialism Imperialism	Corridors and Sea routes	Trade Agreement	Markets
GLOBAL	Oceanic masses	Globalization	Air Transport and Telecommunications	Multilateral Agreements	Interdependency and comparative advantages

FIGURE 5.3 Economic, Social, and Environmental Impacts of Transportation Infrastructure Development (Modified after Jean-Paul Rodrige et al. 2020).

passengers and cargo (Filip Biljecki et al. 2013), which are explained above and also shown in Figure 5.2.

5.1.2 ASPECTS OF TRANSPORTATION DEVELOPMENT

The growth of transport as a system depends on many factors such as level of influence, economy, technological advancements, urbanization, culture, political influence, and environmental consideration (Jean-Paul Rodrige 2020). Figure 5.3 illustrates various aspects to be considered while planning any transport system. Reliable data collection considering all these aspects and its analytics using suitable tools can be helpful in planning, design, construction, and operation of transportation system as well as its services considering the complex correlation between its physical and human needs.

The transport system development enhances the economy and social interactions. This also induced primary, secondary, and tertiary socioeconomic benefits. Such benefits impact the supply and demand of transport, which ultimately improves the system's economy. Benefits of transportation system in terms of primary, secondary, and tertiary. The benefits includes time and cost saving, improves mobility, increases social opportunities, and increases competitiveness (Jean-Paul Rodrigue 2020). It also attracts economic activity, forms a distribution network, and opportunities for education and income source from transport developed and increased, as shown in Figure 5.4.

Primary Benefits: These benefits directly help by optimizing cost and travel time resulting from enhancement in the transport system capacity and efficiency.

Secondary Benefits: These are indirect impacts linked with the improvement of accessibility and economics of the region.

Tertiary Benefits: The benefits are induced due to improvement in the economy, accessibility, and better opportunities for education, social interactions, and leisure.

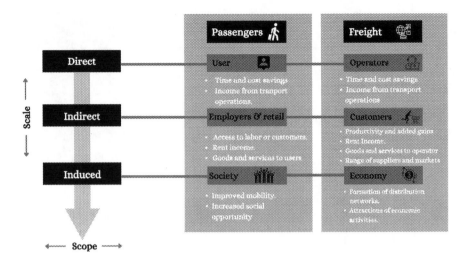

FIGURE 5.4 Transport System Benefits (Modified after Jean-Paul Rodrigue 2020).

5.2   DATA ANALYTICS IN TRANSPORTATION INDUSTRY

The field of transportation engineering is distinct, due to a number of stakeholders participating in the decision-making processes. The decision making applying to the issues faced in transportation engineering requires reliable data for analysis. Before beginning any project planning phase, the transport engineer must obtain the data available for the area under consideration and details of the previous system in place. Planning surveys are conducted to make sound decisions regarding the project's scope, policies, and financing. These surveys assist the engineer in creating a suitable model of the proposed system by precisely estimating the future requirements or circumstances. Planning surveys include physical surveys, economic surveys, and social surveys. This database preparation must take account of collecting data on:

- Governing structure
- Demographic data
- Land use and land cover
- Economic base of the region and financial resources available
- Present transportation services and infrastructure
- Travel patterns and volumes
- Expectations of the community their values, culture, and customs

Transport system planning is a complex entity; therefore, after collecting the suitable data from these sources, analysis to understand the behavior is essential for evaluating the system's performance. The collection of traffic data, its analysis, and suitable application must be understood methodically as part of the planning process. The most important phases in the transport planning process are:

- To finalize the specification of the data required.
- To design the survey methodology.
- To conduct the survey.
- To analyze and validate the data, and
- To apply the data for design, modeling, and implementation.

5.2.1 Significance of Data Analytics in Transportation

The implications of data analytics in transport engineering are multifaceted like any other field, creating a vast reserve of knowledge with meaningful information and valuable data that has the potential to optimize performance and play a crucial role in achieving a strategic edge. Analytics is a broad concept that includes various processes, and analytical methods may be qualitative and objective. It processes raw information collected from various sources into valuable data. Its further analysis resulted in numerical values, patterns, and figures that can support the evidence-based decision-making process for the strategic growth of the transport system by forecasting and better understanding the customer choice, behavior, and future trends. The increase in population has caused a significant increase in the number of private vehicles on the roads, leading to congestion, emissions, pollution, and accidents. But in recent times, emerging technologies such as IoT, big data, data analytics, artificial intelligence, machine learning, and their applications in the projects like smart cities, intelligent transport, multi-modal transport planning, and transportation management are transforming and powering analytics in the field of transport engineering. It is projected that rapid and unprecedented growth of transportation analytics may reach $27.4 billion globally by 2024. The application of data analytics in the transportation offers following advantages:

- The transformation of the transportation segment as a whole.
- Availability of real-time data and information is increasing the system's efficiency.
- It offers flexibility to all the stakeholders.
- Ensuring safety and avoiding accidents.
- Reduced consumption of fuel.
- Better connectivity and inner-modal planning ensure customer satisfaction.
- Predict future trends.
- All the stakeholders in the industry are well informed and can choose the better product, services, etc.
- The operators used data to identify the possible shortage in supply, peak hours consumption of fuel, real-time route optimization, parking slot identification, etc.
- The service providers get a better competitive advantage over the others.
- It also supports environmental intelligence by providing data on pollution, traffic density, noise, etc., variac using sensors.

As transportation planners need to understand the importance of the data before planning any project. For instance, the Pune Municipal corporation is contemplating providing a continuous cycle track throughout the limits of its corporation. If the cycle tracks are constructed without considering whether or not there is a demand for it, then it will lead to two possible outcomes:

- The PMC might end up spending huge amounts of money to construct cycle tracks, and people will not use it since there is no demand for it. That will be a huge waste of money.
- The PMC will spend a tiny amount of money on building cycle tracks in limited areas, but it might so happen that there is huge demand for cycle tracks and therefore the PMC would end up not provide enough infrastructure as per the demand.

Therefore, it is important to understand the demand of any infrastructure before planning to construct it.

For building a better rail network different data required such as geographical, population in area, political data, and educational, industrial data. For data collection, analyzing is most important to prepare predictive models and machine learning uncovered data correlation. The rail network improves the safety, reliability, efficiency, sustainability and network health, as shown in Figure 5.5.

5.2.2 DATA COLLECTION TOOLS AND METHODS

Over the past few decades, the development in the information technology sector has taken all branches of engineering forward by leaps and bounds. Apart from data collected by various physical surveys, presently, an enormous quantity of data is accumulated regarding the movement of the people and goods at record scales through GPS, smartphones, tablets, traffic sensors, imaging devices, and other emerging tools during transportation infrastructure testing, monitoring, and control.

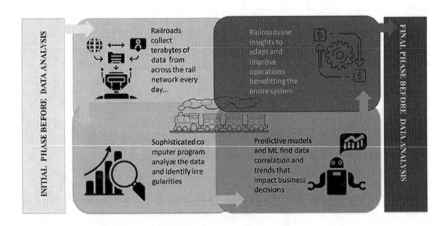

FIGURE 5.5 Building a Better Rail Network Using Data Analysis.

Data is continuously emerging and thus creating a gold mine of data that can be analyzed and utilized to generate multiple predictions. Data is regarding the usage of different transportation modes, movement of people, spatial and temporal interaction of goods and people movement, travel preferences, and frequency of the passengers, variation in the traffic flow patterns, and much more. The availability of such a gigantic volume of data and information makes it more complex to take a suitable conclusion. To process, this gigantic data available to the tune of petabytes requires emerging technologies, advanced techniques, algorithm, probabilistic, and statistical techniques. Also, there are challenges related to storage, computational power, data mining, and conclusions while handling such a huge quantity of data in the case of essential infra projects. To overcome such challenges, the following steps can be followed:

- Prepare the standard format for data collection.
- Develop a suitable model.
- Reorganize the data into various clusters, categories and classes.
- Data integration.
- Uncover hidden relationships, patterns, and correlations.
- Interpretation and conclusions.

5.2.3 PROCESS OF DATA COLLECTION (SURVEYS)

Transportation planning surveys are conducted using suitable tools to collect appropriate data for sustainable planning and development of inaccessible and backward areas. Surveys are conducted for collecting of a region data to understand their vital problems resulting from a lack of accessibility from transportation facilities, services, and networks. All the methodologies used in surveys, proceedings, and stages are essential for authentic data collection. The survey stages are depicted in Figure 5.6. Surveyors should follow these stages, and methodologies for obtaining reliable information and data for successfully conducting survey work (Richardson et al. 1995).

5.2.4 DATA COLLECTION TECHNIQUES

The inception of ICT has transformed the overall picture for the development of all the sectors, including the transport system. This new trend has also upgraded the transportation survey techniques for data collection (Fricker, 2005; Zhang, 2000). These modern survey methods can support the growth in remote regions, which seems inescapable for the sustainable development of the transportation sector (Leeuw, 2005).

5.2.4.1 Traditional Data Collection Techniques

Traditional surveys have been utilized over the years for data collection. These surveys can be bifurcated into two main categories, i.e., supply-side and demand-side surveys. After finalizing the aims, objectives, and duration for the survey, the next step is to choose a suitable data collection technique for conducting the cross-sectional or

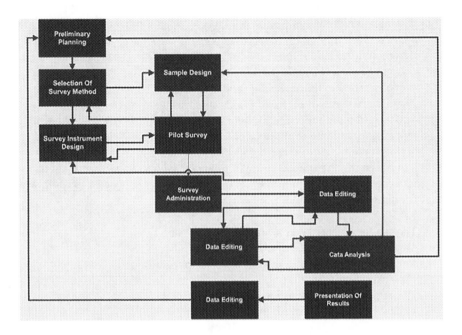

FIGURE 5.6   The Transportation Survey Process (Modified after Richardson et al., 1995).

longitudinal survey. In the case of a longitudinal survey, it must be noted that the same technique is to be used for all the waves of the longitudinal survey; otherwise, conclusions from the data may be erroneous. Following are the traditional data collection techniques (Richardson et al. 1995):

- Documentary searches
- Observational surveys
- Household self-completion surveys
- Telephone surveys
- Intercept surveys
- Household personal interview surveys
- Group surveys
- In-depth surveys
- Questionnaire surveys
- Indirect estimation

In *documentary searches*, the survey aims to collect documents without any response. The *observational surveys* are conducted for gathering the inanimate or animate data without expecting any particular response. In self-completion surveys, it is necessary to get responses and participation in the survey. The interviewer conducts the telephone survey without face-to-face contact. In the Intercept survey, respondents stopped for an interview while using a mode of transport or performing an activity. Following specific information related to trip data, land use data, employment, and general activities are collected from these surveys (modeling Transport by Ortuzarr). 8

- Infrastructure and existing services inventories.
- Land use inventories.
- O-D travel surveys and associated traffic count.
- Socioeconomic information.

The *questionnaire* can include details the planners need to design any infrastructure work. It can include questions based on the respondents' travel like destination, the mode of transport they prefer, the time and cost they need for transport, etc. Such data is then analyzed to understand what factors affect the behavior of the respondents concerning their choice and type of transportation. The analysis is done to understand the demand for a specific infrastructure project and the shortcoming of the current infrastructure.

Indirect estimations are a "quick fix" to draw conclusions or estimate from factors that might not be directly related to our investigations. To put it more succinctly with an example, consider the toll collection at toll plazas. The total amount collected from a toll can estimate the number of vehicles passing through the toll. Even though the toll plaza does not collect this data for transportation planning, transportation planners can make use of this data to generate a quick estimate of the traffic. Since the purpose of the toll plaza is not to collect data for transportation planning purposes, this data becomes "indirect" data, and the subsequent estimation becomes "indirect estimation." However, sometimes direct and indirect data can be combined to generate a decent and near-accurate estimate. Thus, data collection and analysis are an important and integral part of the transportation sector.

5.2.4.2 Modern Data Collection Techniques

Conventionally, travel surveys were performed through face-to-face interviews, intercept surveys, or other techniques. The data collected from such surveys were used as a sound base for developing first-generation travel forecasting models. Considerable changes in the lifestyles of urbanites made the use of conventional traditional survey methods extremely costly and time-consuming. In the last few years, various advanced technologies have been evolved to create new opportunities and transformed the process of conducting surveys for transportation system planning. These techniques are the best fit to overcome the challenges planners face due to changes in the traveler's lifestyle and application of various ICT tools such as computers, electronic data processing (EDP) tools, etc., resulting in efficiency and cost-effectiveness. A few methods of data collection using ICT tools are discussed below:

- **COMPUTER-ASSISTED DATA COLLECTION**

Computers can be used to assist observation and interviews as a means of data collection, and various parameters of travel patterns are assessed with accuracy and constancy compared to traditional data collection tools. The right tool for establishing movement data may depend on the objectives and characteristics of a particular survey, either computer-assisted observation or interviewing. These techniques collect data of better quality due to the exclusion of causes of errors that arise during the manual intervention in data feeding and coding.

1. **Computer-Assisted Observing**
   Automatic positioning techniques such as global system for mobile communication or global positioning system are applied for the traveler's real-time and space data collection.
2. **Computer-Assisted Interviewing**
   The computer-assisted interviewing methods can be broken down into the following categories (Ettema, 1996), (Leeuw, 2005).
   • Computer-Assisted Telephone Interviewing
   • Computer-Assisted Personal Interviewing
   • Computer-Assisted Self Interviewing
3. **Application of Drones/Cameras for Traffic Data Count**
   The real-time traffic data of a particular locality can be video-graphed using various advanced cameras and drones. Further, by using video data analytics, required data can be extracted in terms of traffic volume count.
4. **Application of Monitoring, Image Processing, and Remote-Sensing Technologies**
   The advances in automatic monitoring and remote-sensing technologies, like global positioning systems (GPS) and vehicle instrumentation, when linked to well-developed GIS databases, will provide substantial opportunities to enhance the amount, detail, and exactness of the data collected in twenty-first-century travel surveys. Image processing techniques are also applied for traffic counting and further classifying the data.
5. **Application of the Internet and Multimedia Methods**
   Availability of advanced tools on the Internet and the advent of multimedia computing techniques can evolve better surveying methods for planning quality service to the prospective users.
6. **Combination of Tools to Design Survey Methods**
   Combining more than one tool, e.g., blending telephone, e-mail, and personal interview techniques for personal/household data collection, can be implemented for optimizing the time and cost of data collection.
7. **Automatic Traffic Counters (ATC)**
   There are two types of ATCs used for traffic data collection: fixed type and portable type. Many technologies such as pneumatic road tubes, piezo-electric sensors, and passive infrared devices are employed in designing such counters.

5.2.5 ADVANTAGES OF PREDICTIVE ANALYTICS FOR PUBLIC TRANSPORTATION PLANNING

Nowadays, public transportation is a major part of many people's lives. The transportation agencies face the challenges of reducing costs, increasing returns, and trying to limit or act on time when unexpected events, delays, or accidents happen. Besides, transportation agencies must consider the needs and expectations of travelers. The delay of a mode of transportation due to traffic congestion in general often affects travelers' plans. Big data analytics and predictive analytics come into the picture to resolve all these things.

The predictive analytics applications in the transportation industry were widely discussed during the 100th annual TRB meeting in 2021. This year, traffic management became highly challenging. There is a variation in the peak hours, peak periods, week day and weekend traffic. Also, some areas have a different rebound. As reported by one of the speakers at TRB, Anita Vandervalk-Ostrander, the industry requires new, more effective data sources and processing methodologies. The experts from transportation should focus on the different data streams that are coming from:

- Smartphone data
- Real-time passenger loads
- Traffic flow
- Ridership detailing
- Shared mobility data (Uber, Lyft, etc.)
- National transit database
- Real-time or near real-time counts
- Telework data analysis and reporting

This predictive analytics has following applications:

1. Solving promptly traffic congestion and issues: Predictive analytics can be used in accidents or unpredictable situations such as natural disasters to understand where and when traffic congestion will be highest. Transportation agencies can grip real-time data to apply solutions that alleviate traffic congestion, such as giving additional services during peak hours or providing alternative routes to avoid congestion. Transportation agencies can redirect traffic or put out speed limits in bad weather that avoid or reduce vehicle accidents. In Ireland, in partnership with IBM, Dublin has harnessed big data, using tools such as sensors, traffic detectors, GPS, and CCTV, to find out and resolve the leading causes of traffic congestion in the public bus network.

2. Increase revenues and improve operating costs: Data shows lots of fascinating facts to increase revenue and optimize costs, such as the peak days and peak times of traffic, the number of passengers taking public transportation or using specific routes, real-time traffic situations, etc. Due to this, planners can manage their services, reduce costs, and guarantee a punctual and reliable service in a better way, especially during peak periods. With more advanced algorithms, planners can analyze different traffic scenarios during accidents, local events, or natural disasters. This allows operational planning procedures to ensure traffic flow by controlling and offering alternative routes to avoid disruption in services.

3. Planning efficiently urban/developments projects: Predictive analytics can be used for sustainability projects and ensure smooth and safe mobility for all travelers. This allows planning the impact of urban development projects on the transportation industry. Public transportation systems can benefit from predictive analytics. Data sources and reputable providers are not challenging to locate. Governments, transportation agencies, and urban planners can make more informed decisions by identifying and implementing efficient methods and tools to analyze, exploit, and process combined datasets for reliable forecasting.

5.3 APPLICATIONS OF DATA ANALYTICS IN TRANSPORTATION PLANNING AND MANAGEMENT

In the last few decades, due to the progress of ICT, enormous data related to the movement of goods and people from one place to another have been collected using advanced data collection technologies and techniques to find data sets suitable for traffic planning challenges by means of a variety of tools such as GPS, social media platforms, and compatible applications on mobile devices such as mobile phones and tablets and sensors. In such projects, traffic is highly pretentious by factors that cause high uncertainty due to drivers' or travelers' preferences. For this reason, techniques such as fuzzy sets, metaheuristics, neural networks, and Internet of Things, etc. are most likely used by the research community. Following are few examples of application of data analytics in transportation system planning and management.

5.3.1 DATA ANALYTICS FOR PLANNING THE MULTIMODAL TRANSPORTATION

Presently, a multimodal transportation system has been adapted to ensure flawless travel by integrating different modes of transportation, such as rail, bus, bicycle, tram, metro, railway, airport, waterways, walking, and even personalized vehicles. This network can integrate various modes of public transport and offer end-to-end transportation connectivity to the citizens. The successful integration of the routes is likely to improve the catchment area and the subsequent use of public transport and the efficiency of public transport by reducing the need for supply bus services and the overall need for bicycles. Implementation and planning of a multimodal transportation system necessitate systematic data collection through advanced technological tools such as cameras, GPS, and geo-location. The unmatched amount of data collected using these tools can be analyzed using advanced analytical techniques to augment functioning, optimize costs, and better service to travelers by multiple providers in the transportation industry. The predictive analytics of live data can assist in the following ways to the commuters or transport agencies (Figure 5.7).

- To choose the best connectivity network for origin and destination of trips.
- To get the "best possible result" using live data as a substitute of historical data and information for predictive analysis.

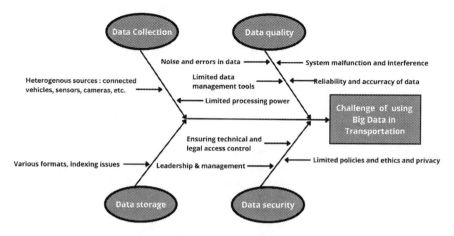

FIGURE 5.7 Structure for Data Analysis.

- To understand and carry out the impact analysis for public transport clo-sures due to unprecedented events like strikes.
- To identify and predict traffic density, accidents, or vehicle breakdown and suggest suitable action.
- To carry out the impact analysis of various transportation infrastructure projects and find an alternative plan without hampering present mobility.

5.3.1.1 Examples of Cities with MTS

Booz Allen, 2012 in his report, described the following (refer Table 5.1) examples of cities, provided with integrated multi-modal transport network using data ana-lytics. Table 5.2 depicts that the success of integrated multi-modal transportation in these cities results from the application of advanced ICT tools like IoT, artificial intelligence, and data analytics in the planning, implementation, and operation of the transport network. To encourage the traveler to use multimodal transport, integration of land use and physical infrastructure is also required.

A few Indian cities are also trying to build multimodal transport systems by integrating metro rails, bus rapid transit systems (BRTS), and other modes in cities like Ahmadabad, Mumbai, Pune, Bengaluru, Hyderabad, etc. The following steps are taken in India to integrate various modes using ICT-based tools (Figure 5.8):

- National Transport Card has been proposed for integration of the fare collection.
- Sharing of cycles by installing bicycle stands near BRT stops and metro stations.
- Grouping of private bus operators in Delhi for cohesive bus networks into 17 clusters and allocating each cluster to one provider.
- Integration of mobile information and internet technology to provide real-time information to bus users.

TABLE 5.1
Case Studies of MTS

City	Features of the MTS
London	• Rail network integrated with bus network and ferry network. • Stations are designed to handle interchange for high volumes of traffic. • Bus exchange routes are built between train and subway stations.
Hong Kong	• Integration of rail network, city buses, trams, taxis, minibusses, and ferries resulted in a high share (90%) of public transport and vehicle ownership rates are very less (50 vehicles per 1,000 population). • MRT system with 11 lines covering 230 km and connected with bus stops, airport, and ferry terminal. • At Sheung Wan MRT station (underground) lies below the Hong Kong-Macau ferry terminal, and two bus stations are nearby. • Even though several operators provide services, good integration, coordination, and connectivity resulted in the availability of transport mode within walking distance of six minutes only. integration, coordination, and connectivity resulted into availability of transport mode within walking distance of six minutes only.
Singapore	• Well-established integrated multi-modal transport network with strategically located transit stations. • Licensing of an area, electronically pricing the road system, and rationing system to limit vehicle ownership.
Bogotá, Columbia	• Integration is achieved through 'feed routes' (309 km) connected by dedicated bus routes (84 km). • Bicycle lanes and pedestrian paths (pedestrian plazas without cars) connect to the BRT network. • Stations are available to reduce travel distances and integrate land use planning for high density, mixed land use near bus stops and intersections. • Demand control through measures to prevent the growth of private vehicles, reduction of parking capacity, and restriction of access to the number of vehicles entering the city by preventing vehicles with certain number plates on certain days. • Integration of fares and institutional changes by coordinating and integrating all providers.

5.3.2 FORECASTING THE TRAFFIC CONGESTION

Today, commuters choose a car as the preferred mode of transportation over urban public transport. In the cities, the number of vehicles per capita has increased in the last ten years, and traffic congestion has lowered the efficiency of the entire transportation system. Due to that, there is a need to invest in transportation research to improve current transportation systems. According to Antonio D. Masegosa et al. (2018), the economic costs of traffic congestion would rise by 50% by 2050, the accessibility difference between critical and peripheral places will deepen, and unplanned and pollutants-related socioeconomic charges will continue to rise.

TABLE 5.2
Application of Data Science and IoT for Multimodal Transportation

Application	Application of Data Science and IoT
Smart Cards for Integration of Fare	• Combined fare collection using Oyster Card (London), Octopus Card (Hongkong), and EZ Card (Singapore), which applies to all types of public transportation, parking, and shopping at small shops. • Commuters **may** pay **with** contactless ards or **portable** wallets. • In Singapore, data collected from fare cards and sensors are used to track buses. Further, this data is analyzed for transport optimization.
Smart Transport System for Integration of Information	• Good public transport signage and real-time information systems (e.g., time for next bus to arrive). • The TransitLink guide and a number of signs provide complete information on all aspects of travel. • Singapore city planned the "Smart Mobility 2020 initiative" for providing real-time travel information to the commuters using ITS. This project aims to distribute morning peak hour traffic evenly by encouraging users to re-plan their journey based on crowd data. • Vehicle tracking.
Operators' Integration	Single governing authority or multiple operators but their coordination using ICT tools leads to encouraging the travelers to use public transport.
Integration of the Routes	Providing interchange points for safe and convenient transfer of the passengers from one mode to another.
Data Collection	To plan and implement an integrated multimodal transport system for a city, reliable data is essential. Data collection is a costly affair if conventional methods are used. However, using tools like mobile phones and smart cards, data can be generated comparatively easily.
AI Application	Real-time performance management, customer statistics, predictable network planning, and route construction. Integrating AI with IoT and blockchain can lead to intelligent channels, customer sensor tracking, customer data engine, event simulation, disruption, and an intelligent urban grid.

In this field, the following techniques are getting used for the previous few years with a high impact and reliable performance to forecast the traffic condition on the highway and concrete roads. The primary venture in this field is to predict on a time shot using data analytics with the possibility of reducing congestion, reducing journey time, reducing CO_2 emissions, less consumption of the fuel, and decrease in noise pollution on the urban roads and highways (Figure 5.9).

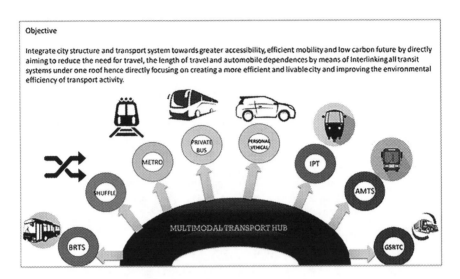

FIGURE 5.8 Multimodal Transport Hub.

FIGURE 5.9 Traffic Congestion.

- **Prediction of Traffic Congestion Using Neural Networks:** Literature review revealed that the application of neural networks for the prediction of traffic congestion has increased in the last few years. A neural network is used to predict future traffic flow values, depending upon the current value and aggregate past values. In few models, traffic congestion anticipated the usage of beyond site visitors facts such as quantity, pace, density, time of the day, and particular day of the week used as input variables. The deep belief network (DBN) was created to improve the link between meteorological data and traffic flow. This is used to improve the accuracy traffic flow forecast during adverse weather conditions (soft computing based optimization and decision models, Springer Science and Business Media LLC 2018).

- **Prediction of Traffic Congestion Using Fuzzy Systems:** This system allows knowledge-based total analysis used for efficient traffic congestion detection. The proposed fuzzy inference model takes flow and density of the traffic as inputs, and the output is discovered in terms of congestion ranges (ranging from "no congestion" to "severe congestion") suggests in their research that this model for congestion forecasting might be most suitable transport planning, control, and risk assessment (Kalinic and Krisp, 2019).
- **Vehicle Route Planning Using Soft Computing:** Vehicle route planning is one of the most important and frequent activities in transport management. The development of information systems in route systems faces the challenges of planning complexity, lack of accuracy, and completeness of data. This study is used to solve the close-open vehicle routing problem by applying fuzzy optimization approach with the help of hybrid CO metaheuristic procedure to solve the problem with an optimum solution (J. Brito et al. 2015).

5.3.3 SELF-DRIVING CAR

By introducing self-driving cars and offering a safe driving environment for drivers, data science has established a firm foothold in the transportation business. It is also helping in optimizing the performance of the vehicle by analyzing fuel consumption, driver's behavior, regular vehicle monitoring, and providing superior autonomy to the drivers. Various parameters like consumer behavior, geographical position, economic factors, and logistic details can guide the vendor in finalizing the route choice and optimizing the resource allocation (Thakurdesai et al. 2021).

Using data, it was possible to develop machine learning algorithms that master *perception, localization, prediction, planning, and control.* The machine learning algorithms also ensure that the cars learn how to perform the suitable task as good as (or even better than) humans. Perception uses a variety of sensors to determine the location of the road and the state of each obstacle (type, position, and speed). Localization relies on very specialized maps and sensors to comprehend where the car is in its environment at a centimeter level. The car's ability to predict the behavior of objects in its environment is known as a prediction. Knowing the car's position and barriers are used to plan routes to a destination. The law's application is coded here, and the algorithms specify waypoints. The goal of control is to create algorithms that efficiently follow the waypoints (Figure 5.10).

A self-driven car is equipped with features like precise geographical maps of road, sensors, cameras, LIDAR, vehicle-to-vehicle cloud communication system, and sensor date input to the car's machine learning algorithms for forecasting and planning, and suitable action using the vast volume of data. It is anticipated that each self-driving car could collect up to 1 GB of data per second. It also requires a number of sensors and a robust system for immediate and constant transmission of information (Figure 5.11).

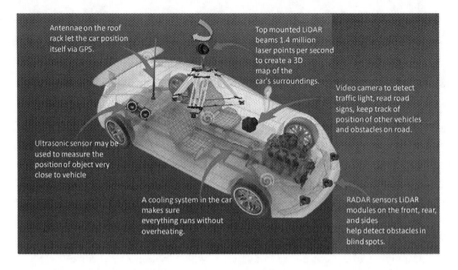

FIGURE 5.10   Model of Self-Driving Car (Modified after Thakurdesai et al. 2021).

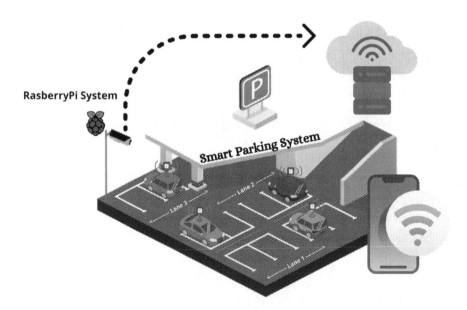

FIGURE 5.11   Smart Parking System (Modified after Khanna et al. 2016).

5.3.4 FINDING PARKING SLOTS

This is an IoT-based cloud-integrated smart parking system, and is comprised of an on-site IoT module that monitors and sends real-time signals on parking slot availability developed by (Khanna and Anand, 2016 & Jung HG et al. 2006). The mobile application is also developed that provides information about the availability of parking slots, and accordingly, it can be booked by users from remote locations.

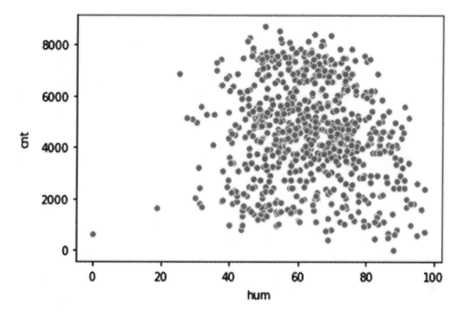

FIGURE 5.12 Scatter Plot of the 'cnt' Variable and 'hum' Variable.

This study also describes the proposed system architecture, mathematical model, and its workings. The smart parking system is represented in Figure 5.12. The components of the system are as follows:

- Sensors: Many sensors, including infrared, passive infrared (PIR), and ultrasonic sensors, are linked to a 5V supply via the Raspberry Pi or an external source. The processing unit consists of a Raspberry Pi that acts as a processor on-chip and connects the cloud to the sensors.
- Mobile application: This application worked as an intermediate to the users for interacting with the system.
- *Cloud*: A server is hosted on the cloud. The database related to parking records, such as the timing of parking, total duration, amount paid, mode of payment, etc., are stored on the cloud, and users can access the system.

In this work, a parking automation system based on a programmable logic controller (PLC) with supervisory control and data acquisition was proposed (SCADA). The modeling of the proposed system was also described in the study. This system provides instructions for parking the car in a specific pattern and relevant security mechanisms.

The suggested system provides safety, parking guidance based on specified norms, efficient parking space utilization, and parked vehicle security. The car's security has been prioritized in this system, as the car can only be removed using an exclusive passcode once it has been parked. The developed system, which operates in real time, can reduce parking issues and provide clients with a high level of security.

5.3.5   THE CONNECTED VEHICLE

The Internet of Things (IoT) is a network of interconnected nerves that can supply city planners with valuable data. With the help of intelligent programs, IoT data can be used to optimize traffic flow and service performance (Mehmet Fatih Tuysuz et al. (2020).

Devices connected to IoT include smartphones and other personal devices that most of us use and carry daily. Planners and other authorities have a lot to gain from these connected devices. The linked car environment is a significant example, even though it is still in its early stages. Driving can be made safer and more efficient with the help of connected technologies. There are five possible connected car models (Paiva et al. 2021):

- **Vehicle to infrastructure:** This technique connects the vehicle to traffic infrastructure, which can convey information about road conditions and the likelihood of delays or accidents.
- **Vehicle to vehicle:** Data can be transmitted back and forth between two vehicles to evade accidents and avoiding congestion.
- **Vehicle to cloud:** Connecting a car to the cloud enables the speedier dissemination of a variety of information, such as traffic conditions and alternative routes, as well as information about gas stations, available resources, and retail places.
- **Pedestrian car:** According to some experts, automobiles, and trucks will be able to speak with walkers via their smartphones shortly. This holds a lot of promise in terms of enhancing pedestrian safety. It can also provide new avenues for mobility.
- **Vehicle to everything:** Over time, a holistic connectivity paradigm in which vehicles are connected to all of the above will most certainly arise.

5.3.5.1   Improved Traffic Flow Using Technology

Engineers are already putting connected technology to work to enhance traffic flow, mainly by extending the usage of the next generation of intelligent robots. Using traffic control systems, the lights at traffic signals can alter at peak times by modifying their sequence and phase times to fulfill specific needs. The lights on high-volume roadways remain green for a long time, rather than employing preset colors that increase the congestion and distraction.

5.3.5.2   Smart Transportation for Intelligent City Planning

The "smart city" movement is gaining hold worldwide, giving urban planners a whole new set of options. Smart cities use an information and communication technology framework to implement and drive development approaches. They build connected technology-equipped and appropriate infrastructure fit for the future.

By establishing intelligent and unique transportation models, smart transportation has the excellent goal of alleviating challenges. Smart transportation is a framework that uses the power of ICT for the acquisition, management, and mining of traffic-related data. Sources, which, in this study, are categorized into 1) traffic flow sensor; 2) video image processor; 3) person and vehicle exploration using

GPS, cellular, and Bluetooth networks; 4) location-based social networking; 5) data transmission smart card–focused data; and 6) environmental data. The ultimate efficacy of innovative transportation architectures is highly dependent on the quality of data collected and features of the analytical techniques used for data analysis.

It all starts at the city planning level, with city leaders committing financially and philosophically to develop a smart city. Effective SLM-driven traffic regulations include traffic congestion, much-needed parking, air leakage, and toll roads on highways and highways. Congestion pricing is a system in which drivers and users of public transportation pay higher rates to access roads and services during rush hour. It makes it impossible for people to join traffic during rush hour, reducing traffic congestion and flow. On the other hand, parking prices are determined by demand and are designed to maximize both local use and availability.

In European cities, low-emission zones are becoming more widespread. They are intended to reduce pollution and improve local air quality by limiting or prohibiting the admission of high-polluting cars into low-emission metropolitan zones. Toll lanes are becoming more common in North America and other parts of the world. There are a variety of payment schemes in existence, with one of the most frequent requiring motorists to pay to use the roadways. Cars carrying a specified number of passengers may be exempt from the fee, providing a financial incentive for carpooling. Some cities, for example, use traffic statistics to limit travel to specific days, while others use traffic jams to alleviate traffic congestion and congestion.

5.4 BOOM BIKE-SHARING DEMAND CASE STUDY

Now that we know the types of data, its sources, and its uses in transportation engineering, we will now understand with the help of a case study the practical field application of data analytics in the transportation engineering sector. We will look at a very famous case study called the "Boom Bikes demand analysis" or the "Boom Bike Sharing," which is frequently used to understand regression applications in the transportation sector. Let us first look at the problem statement and then answer the questions based on the case study data set using some Python programming.

5.4.1 PROBLEM STATEMENT

A bike-sharing system is a service in which bikes are made available for shared use to individuals on a short-term basis for a price or free. Many bike share systems allow people to borrow a bike from a "dock," which is usually computer-controlled wherein the user enters the payment information, and the system unlocks it. This bike can then be returned to another dock belonging to the same system.

A U.S. bike-sharing provider, **Boom Bikes**, has recently suffered considerable dips in its revenues due to the ongoing coronavirus pandemic. The company is finding it very difficult to sustain itself in the current market scenario. So, it has decided to come up with a mindful business plan to accelerate its revenue as soon as the ongoing lockdown comes to an end and the economy restores to a healthy state.

In such an attempt, Boom Bikes aspires to understand the demand for shared bikes among the people after this ongoing quarantine situation ends across the nation due to COVID-19. They have planned to prepare themselves to cater to the people's needs once the situation gets better all-around and stand out from other service providers and make huge profits.

They have contracted a consulting company to understand the factors on which the demand for these shared bikes depends. Specifically, they want to understand the factors affecting the demand for these shared bikes in the American market. The company wants to know:

- Which variables are significant in predicting the demand for shared bikes?
- How well do those variables describe the bike demands?

Based on various meteorological surveys and people's styles, the service provider firm has gathered a large dataset on daily bike demands across the American market based on some factors.

Business Goal: It would be best to model the demand for shared bikes with the available independent variables. The management will use it to understand how the demands vary with different features. Accordingly, they can manipulate the business strategy to meet the demand levels and meet the customer's expectations. Further, the model will be a good way for management to understand the demand dynamics of a new market.

5.4.2 ADDITIONAL QUESTIONS

1. What are the significant variables that help predict the demand for the bikes?
2. What are the insignificant variables that do not predict the demand for the bikes?
3. What are the top five variables that the company should focus on to strategize and increase the demand for the bikes?

Notes: The data set and data dictionary is provided in the Appendix.

5.4.3 UNDERSTANDING THE DATA SET AND THE DATA DICTIONARY

The first step to analyzing any data set is to understand the nature, size, and quality of the data. A simple command in python gives the size of the data set. It is essential to know how many data points are to be analyzed. The Boom Bike data set has 730 unique data points with 16 columns for each data point (ref. Figure 5.12).

A data dictionary (see Appendix) helps understand the abbreviations used in the data set. For example, in the case of the current case study, let us look at some examples from the data dictionary:

- Column temp is the temperature in Celsius
- Weather site is the weather situation, and the entries in these columns are 1, 2, 3, 4

where
1. Clear, few clouds, partly cloudy
2. Mist + Cloudy, Mist + Broken clouds, Mist + few clouds, Mist
3. Light Snow, Light Rain + thunderstorm + scattered clouds, Light rain + scattered clouds
4. Heavy rain + Ice Pellets + thunderstorm + mist, snow + fog

A prior understanding of the data set helps familiarize with the variables and the data type (categorical or numerical). This is important to decide which manipulating data techniques are to be used or how to approach the analysis.

5.4.4 SOLUTION APPROACH

Step 1. Loading the Required Libraries
Step 2. Loading the Data Set
Step 3. Investigating the Data Set
Step 4. Exploratory Data Analysis
Step 5. Data Cleaning and Data Preparation
Step 6. Model Building
Step 7. Model Evaluation
Step 8. Iterations of the Model
Step 9. Interpretation, Observation, and Conclusion

Note: The code as a whole is provided in the Appendix. As we discuss the solution steps in the next sections, relevant code syntax and output will be shared as required.

Step 1. Loading the Required Libraries

In any programming language, libraries are a set of predefined functions that can be used whenever needed. To use these functions, the library in which these functions are "stored" need to be first loaded in the code. Only then can these functions be recalled and used. To simply explain this, let's look at a short example: Suppose for each of the following divisions of third-year engineering – A, B, C there is a register to keep record of each student's marks. Now a student named Ram is in Division B. His record of marks can be found only in the register for Division B. If someone looks for Ram's records in Division A's register, they will not find the record of marks of Ram. In this case, let's call "marks" the function and the "register" its library.

It is important to note that the libraries in python can be loaded at any point of the code. The only condition is that the library must be loaded before using the function that we need. Some programmers choose to load the libraries at the beginning of the code, some choose to load them whenever the library is required. Please look at the following syntax that illustrates how to load a library:

```
importpandasaspd
    fromsklearn.linear_modelimportLinearRegression
```

In the first line of code, the pandas library is loaded as pd. This means whenever we need to use the pandas library we will use it by recalling it as 'pd'.

In the second line of code, the function LinearRegression is imported from the library sklearn.linear_model. This means if we need to build a linear regression model, we will first have to import if from the required library and the recall 'LinearRegression' function to get the model.

Note: It is always better to suppress the warnings that the code throws during the compilation because it makes it a tedious-looking output. The following line of code can generally be used to suppress warnings:

```
import warnings
    warnings.filterwarnings('ignore')
```

Step 2. Loading the Data Set

Just like we loaded the libraries, we always have to load the data set. This is a very crucial step without which the code will not progress. This is always done at the beginning. The data set, that can be in a spreadsheet, is assigned to a variable and loaded as shown below:

$$df = pd.\ read\_csv\ ('path\ where\ the\ CSV\ file\ is\ stored/File\ name.\ csv')$$

Here, df is the variable to which the data set is assigned. pd is a function with which the data set can be stored in the form of a data frame in the variable 'df'.

Step 3. Investigating the Data Set

After the data set is loaded, it is a good practice to look at the data points in the data set to understand the data set. This process is called investigating the data set. It is important to know what are the different columns, rows, their meaning and significance, their units (if the data set has numerical columns) and so on. Once these parameters about the data set are known, the methods and manipulations that should be used for the analysis of the data can be decided. Therefore, investigating the data is a very important part of the analysis.

Note: Due to the diverse nature of data set investigation/exploration, the authors have chosen not to include any code syntax in this section. Data exploration methods that are relevant to this case study can be easily found in the commented Python code after the theory section. Each step has been explained in the code using comments.

Step 4. Exploratory Data Analysis

Exploratory data analysis (EDA) is a way to visually investigate and explore data. In the previous step we looked at the numerical or categorical methods of exploring the data. In EDA, visualizations like bar plots, line plots, scatter plots, box plots,

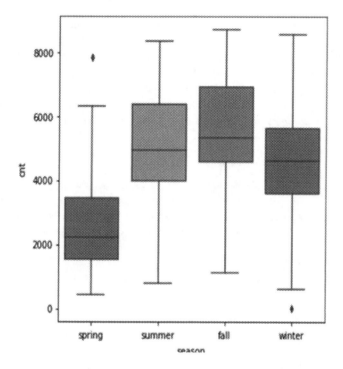

FIGURE 5.13 Box Plots for Nature of Data.

etc. are used to visually represent the data points in our data set. Visualizations help to better grasp the variation, quality, and distribution of the data points in the data set.

Pair plots give a grid of scatter plots where each variable is plotted against each other variable. This gives a quick understanding of the relation of a variable with another variable. Therefore, scatter plots are useful in understanding if there is correlation and how strong it looks, if there is multicolinearity, and so on. For example, Figure 5.12 shows the scatter plot of the 'cnt' variable and 'hum' variable.

Similar to scatter plots, box plots also indicate the nature of the data. More importantly, box plots allow comparison of categorical data. The purpose of the box plot is to visually represent the minimum, the maximum, and the median. A box plot shows the spread of the data as well as the center.

Note: Relevant EDA methods can be found in the code along with comments to understand the rationale behind using them.

Step 5. Data Cleaning and Data Preparation

In the data cleaning step, the data points that are not required or are not relevant to the analysis are deleted (dropped) from the data set to facilitate easy analysis. This step can be performed when sufficient data investigation is done. There is no fixed order to perform this step and can be performed at regular intervals throughout the analysis as wand when it is concluded that particular data points are irrelevant in the further analysis. Data preparation includes preparing/modifying the variables in the data set to be used as per requirement depending on which prediction model that will be used; for example, creation of dummy variables and transforming variables into binary variables are some parts of the data preparation step.

Note: Relevant data cleaning and data preparation methods can be found in the code along with comments to understand the rationale behind using them.

Step 6. Model Building

In this step, the model that has been decided to be used is built to get the desired output. Some examples of models commonly used are simple linear regression, multiple linear regression, logistic regression, and decision trees and so on. In the case study for the Boom Bike share, the multiple linear regression model will be used.

Form of the regression model: $y = \beta_0 + \beta_1 X_1 + \beta_2 X_3 + \ldots.. + \beta_n X_n$

Here:

Y = Dependent variable
β_0 = Constant (y-intercept)
$\beta_{1...n}$ = Coefficient or slope for each explanatory variable
$X_{1 \ldots n}$ = Independent or explanatory variable

Step 7. Model Evaluation

Model evaluation is done to judge the performance of the model. In the case of classification models, parameters like accuracy, sensitivity, and specificity, etc. are used to evaluate the model. In the case of predictive modeling, parameters like R-squared, t-statistic, f-statistic, and p-value, etc. are used to evaluate the model.

In this case study, the parameters used to evaluate the model are R-squared and mean-squared errors primarily.

MSE: Mean-squared error is a metric that is used to check how close or far are the estimates from the actual values. Mean-square error is calculated using the formula:

$$\frac{1}{n} \sum_{i=1}^{n} (Y_i - (Y_i^{\wedge 2}))$$

Where n = number of observations

Y = observed values
$Y^{\wedge}$ = predicted/estimated values

The less the MSE value, the better.

Note: Details of the model and its evaluation can be found in the code. Also, a model, its significance, and interpretation are explained after the code under the discussion section.

Step 8. Iterations of the Model

After the evaluation of the model, there are two possible outcomes: either the model is acceptable or the model need to be improved. If the model is acceptable, then there is no need for further iterations for improving the model. However, if the model needs improvements, further iterations are needed that can include adding or dropping variables, and so on. This step depends on the type and purpose of the model selected.

In the case of the present case study, the variance inflation factors are used to determine the variables that need to be dropped. Variance inflation factor (VIF) is a metric that indicates the strength of multicollinearity of the variables. Multicollinearity is the occurrence of high inter-correlations between two or more independent variables.

A rule of thum to interpret the VIFs is:

1 = not correlated
1 to 5 = moderately correlated
more than 5 = high correlation

Thus, during the iterations, the variables with high multicollinearity are ruled out.

Note that the variables should always be dropped one at a time and the entire process of model building, evaluation, and VIF must be done again. However, for the purposes of this case study and for illustration, the authors have dropped all the variables at once.

Note: Details of the model and its evaluation can be found in the code. Also, the model, its significance, and interpretation are explained after the code under the discussion section.

Step 9. Interpretation, Observation, and Conclusion

This is usually the concluding step of the analysis and results from the final model are interpreted.

A number of parameters are used to interpret a regression model. For the purposes of this case study and the targeted audience, the coefficients and the R-square value are used to interpret the model.

For this case study, in all there were three models that were tried. In the first model, "regressor", the R-squared value is 0.85, and the MSE value is 583842. This, however, is a naïve model that has all the variables. Essentially sticking to this model would mean all the variables are important, which will not be helpful for the company to develop a good strategy. Therefore, with the help of recursive feature elimination (refer code), the count was brought down to the top 20 variables that affect the dependent variable.

The second model, "lm1", is used and the top 20 variables with an R-squared value of 0.834. A post the building of the model, variance inflation factor (VIF), was checked and the variables with VIF values more than five were dropped. These variables were 'workingday', 'season_spring', and 'season_winter'.

The last model, "lm2", thus had 17 variables and the R-squared was 0.805 and the MSE was 789367. While this model has a higher MSE and a lower R-squared, it also has lower variables. Further investigation could also lead to a better model that will have fewer variables and a higher R-squared.

The best model would therefore be a model with optimum number of variables that can be used to develop a strategy to improve the demand for the Boom Bikes with a significantly good R-squared value (> 0.75). Such a model would be the best model.

Two common steps to reduce the number of variables would be to run RFE with a lower variable count or to check the VIF after each model and keep dropping the variables until the best model is built.

Looking at any regression model, the coefficients of the variables indicate the extent to which (magnitude) the independent variable affects the dependent variable. If the sign of the coefficient is positive, then that indicates a positive or a negative correlation between the independent and the dependent variable.

The variables that make up for the final model are:

Holiday, tem, hum, windspeed, season_summer, yr_1, mnth_Dec, mnth_Jan, mnth_ Jul, mnth_Mar, mnth_May, mnth_Oct, mnth_Sep, weekday_Sat, weekday_Sun, weathersit_Light snow, weathersit_Mist.

The model in the form of the equation will be (Figure 5.14):

$$Cnt = 3450.85 - 574.80(Holiday) + 955.64(temp) - 233.85(hum)$$
$$- 298.81(windspeed) + 375.05(season\_summer) + 2046.71(yr\_1)$$
$$+ 49.34(mnth\_Dec) - 820.57(mnth\_Jan) - 282.54(mnth\_Jul)$$
$$+ 80.61(mnth\_Mar) + 195.14(mnth\_May) + 1024.51(mnth\_Oct)$$
$$+ 880.80(mnth\_Sep) + 94.03(weekday\_Sat)$$
$$+ 316.15(weekday\_Sun) - 1806.43(weathersit\_Lightsnow) - 337.$$
$$13(weathersit\_Mist)$$

Finally, taking a look at the graph in Figure 5.15, it can be seen how close the predicted and actual values of y are.

Note: Details of the model and its evaluation can be found in the code.

```
                        OLS Regression Results
========================================================================
Dep. Variable:                  cnt   R-squared:                    0.793
Model:                          OLS   Adj. R-squared:               0.791
Method:               Least Squares   F-statistic:                  321.7
Date:              Wed, 15 Apr 2020   Prob (F-statistic):        1.37e-168
Time:                      19:45:52   Log-Likelihood:              440.46
No. Observations:               510   AIC:                         -866.9
Df Residuals:                   503   BIC:                         -837.3
Df Model:                         6
Covariance Type:          nonrobust
========================================================================
                 coef    std err          t      P>|t|      [0.025      0.975]
------------------------------------------------------------------------
const          0.2596      0.020     12.986      0.000       0.220       0.299
yr             0.2360      0.009     25.713      0.000       0.218       0.254
temp           0.4279      0.028     15.277      0.000       0.373       0.483
windspeed     -0.1524      0.028     -5.536      0.000      -0.206      -0.098
season_spring -0.1413      0.014    -10.427      0.000      -0.168      -0.115
mnth_Jul      -0.0715      0.019     -3.769      0.000      -0.109      -0.034
weathersit_C  -0.2413      0.027     -8.873      0.000      -0.295      -0.188
========================================================================
Omnibus:                     54.941   Durbin-Watson:                1.895
Prob(Omnibus):                0.000   Jarque-Bera (JB):           101.150
Skew:                        -0.657   Prob(JB):                  1.09e-22
Kurtosis:                     4.742   Cond. No.                      10.5
========================================================================

Warnings:
[1] Standard Errors assume that the covariance matrix of the errors is correctly specified.
```

FIGURE 5.14   Results of Model lm2.

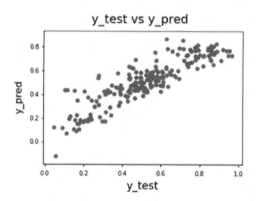

FIGURE 5.15   Graph of Predicted Versus Actual Value of y.

REFERENCES

Booz Allen (2012). "Integrating Australia's transport systems: A strategy for an efficient transport future". Infrastructure Partnership Australia (www.infrastructure.org.au) at www.infrastructure.org.au/DisplayFile.aspx?FileID

Filip Biljecki, Hugo Ledoux, Peter osterom. "Transportation mode-based segmentation and classification of movement trajectories". *International Journal of Geographical Information Science*, Volume 27, 2013, pp. 385–407. 10.1080/13658816.2012.692791.

J. Brito, F.J. Martinez, J.A. Moreno, J.L. Verdegay, "An ACO hybrid metaheuristic for close-open vehicle routing problems with time windows and fuzzy constraints", *Elsevier, Applied soft Computing*, Volume 32, July 2015, pp. 154–163. 10.1016/j.asoc.2015.03.026

E., De Leeuw, & W., Nicholls (1996). 'Technological Innovations in Data Collection: Acceptance, Data Quality and Costs'. *Sociological Research Online*, Vol. 1.

E., De Leeuw (2005). To Mix or Not to Mix? *Data Collection Modes in Surveys. Journal of Official Statistics*, 21, 233–255.

D. Ettema (1996) Activity-Based Travel Demand Modeling. Unpublished Ph.D. dissertation, Faculteit Bouwkunde, Technische Universiteit Eindhoven, 297 p.

Charles Fox. "Data Science for Transport A Self-Study Guide with Computer Exercises", Springer Textbooks in Earth Sciences, Geography and Environment, ISBN 978-3-319-72952-7 ISBN 978-3-319-72953-4 (eBook). 10.1007/978-3-319-72953-4

S., Fricker (2005). An Experimental Comparison of Web and Telephone Surveys. *Public Opinion Quarterly*, 69, 370–39210.1093/poq/nfi027.

H.G. Jung, D.S. Kim, P.J. Yoon, J. Kim, editors. Parking slot markings recognition for automatic parking assist system. 2006 IEEE Intelligent Vehicles Symposium; 2006: IEEE.

Maja, Kalinic, & Krisp, Jukka M. (2019). Fuzzy inference approach in traffic congestion detection. *Annals of GIS*, 25, 329–33610.1080/19475683.2019.1675760.

L.R. Kadiyali, "Traffic Engineering and Transport Planning", L.R. Kadiyali, Khanna Publications 2007.

Maja Kalinic, Jukka M. Krisp, "Fussy inference approach in traffic congestion detection", *Taylor & Francis, Annals of GIS*, Volume 25, issue 4, pp. 329–336, 2019. 10.1080/19475683.2019.1675760

Abhiup Khanna, Rishi Anand. "IoT based Smart Parking System". University of Petroleum & Energy Studies, January 2016. 10.1109/IOTA.2016.7562735

Edith De Leeuw, William Nicholls, "Technological innovations in data collection: acceptance, data quality and costs", *SAGE Journals*, Volume 1, issue 4, pp. 23–37, 1996. 10.5153/sro.50

Lei Lin, Advisors. "Data Science Applicationin Intelligent Transportation systems: An integrative Approach for Border Delay Predictionand Traffic Accident Analysis", State University of New York at Buffalo. ProQuest Dissertations Publishing, 2015. 3683052.

Antonio D. Masegosa, Enrique Onieva, Pedro Lopez-Garcia, Eneko Osaba. "Chapter 4 Applications of Soft Computing in Intelligent Transportation Systems", Springer Science and Business Media LLC, 2018. Gursagar Singh, Patiala SangeetaKamboj. "Modelling of Smart Car Parking System using PLC" E International Journal of Mechanical Engineering and Technology.

Natalia Nesterova, Vladimir Anisimov, "Effective Strategies for Multimodal Transportation Network" *XII International Scientific Conference on Agricultural Machinery Industry IOP Conf. Series: Earth and Environmental Science*, Volume 403, 2019, pp. 012204. IOP Publishing. 10.1088/1755-1315/403/1/012204

John S. Niles. "Telecommunications Substitution for Transportation". *Transportation Engineering and Planning*, Volume II, Global Telematics, Seattle, Washington, USA.

Sara Paiva, Mohd Ahad, Gautami Tripathi, Noushaba Feroz, Gabriella Casalino. "Enabling technologies for urban smart mobility: recent trends, opportunities and challenges" *Sensors*, Volume 21, 2021, pp. 2143. 10.3390/s21062143.

A. J., Richardson, E. S., Ampt, & A. H., Meyburg (1995). Survey Methods for Transport Planning. *University of Melbourne, Parkville*.

Anthony J. Richardson, Elizabeth S. Ampt, Arnim H. Meyburg. "Survey method for transportation planning xiv, 459, 33 pages: illustrations; 24" ISBN:064621439X 9780646214399, OCLC No. 32404041, xiv, 459, 33 pages.

Chaitanya Rindhe, "Smart Car Parking System using IR Sensor" *International Journal for Research in Applied Science and Engineering Technology*.

Jean-Paul Rodrigue. (2020) "The Geography of Transport Systems" Routledge, 456 pages. ISBN 978-0-367-36463-2, New York. 10.4324/9780429346323

Gursagar Singh, Patiala Sangeeta Kamboj. "Modelling of smart car parking system using PLC" *E International Journal of Mechanical Engineering and Technology. (IJMET)*, Volume 9, Issue 7, July 2018, pp. 909–915, Article ID: IJMET_09_07_098.

"Soft Computing Based Optimization and Decision Models", Springer Science and Business Media LLC 2018.

Mir AftabHussain Talpur, Madzlan Napiah, Imtiaz Chandio, Shabir Khahro. Transportation Planning Survey Methodologies for the Proposed Study of Physical and Socio-economic Development of Deprived Rural Regions: A Review. *Modern Applied Science*, Volume 6, 2012, pp. 1–16. 10.5539/mas.v6n7p1.

Hrishikesh Thakurdesai, Jagannath Aghav. "Autonomous Cars: Technical Challenges and a Solution to Blind Spot". 2021. 10.1007/978-981-15-1275-9_44.

Transportation Analysis Market, Report Code: TC 7390, Source: Markets and Market Analysis, Oct 2019.

Mehmet Fatih, Tuysuz, & Ramona, Trestian (2020). From serendipity to sustainable green IoT: Technical, industrial and political perspective. *Computer Networks*, 182, 10746910.1016/j.comnet.2020.107469.

Mehmet Fatih Tuysuz, Ramona Trestian. "From serendipity to sustainable green IoT: Technical, industrial and political perspective", *Computer Networks*. 2020.

Simon P. Washington, Matthew G. Karlaftis, Fred L. Mannering. "statistics and econometric methods for transportation data analysis" second edition, CRC Press, Taylor & Francis group, A Chapman and Hall Book.

Y., Zhang (2000). Using the Internet for survey research: A case study. *Journal of the American Society for Information Science*, 51, 57–6810.1002/(ISSN)1097-4571.

Jinhua Zhao, Adam Rahbee, Nigel Wilson. "Estimating a Rail Passenger Trip Origin-Destination Matrix Using Automatic Data Collection Systems" *Computer-Aided Civil and Infrastructure Engineering*, Volume 22, issue 5. 10.1111/j.1467-8667.2007.00494.x

Appendix –

1. Data Dictionary –

12/25/2020
https://cdn.upgrad.com/uploads/production/0fdfa494-a9bf-4b32-873f-900e3b262948/Readme.tx
Datasetcharacteristics
day.csvhavethefollowingfields:

- instant:recordindex
- dteday:date
- season:season(1:spring,2:summer,3:fall,4:winter)
- yr:year(0:2018,1:2019)
- mnth:month(1to12)
- holiday:weatherdayisaholidayornot(extractedfrom http://dchr.dc.gov/page/holiday-schedule)
- weekday:dayoftheweek
  workingday:ifdayisneitherweekendnorholidayis1,otherwiseis0.
  +weathersit

- 1:Clear,Fewclouds,Partlycloudy,Partlycloudy
- 2:Mist+Cloudy,Mist+Brokenclouds,Mist+Fewclouds,Mist
- 3:LightSnow,LightRain+Thunderstorm+Scatteredclouds,LightRain+Scatteredclouds
- 4:HeavyRain+IcePallets+Thunderstorm+Mist,Snow+Fog
- temp:temperatureinCelsius
- atemp:feelingtemperatureinCelsius
- hum:humidity
- windspeed:windspeed
- casual: count ofcasualusers
- registered:countofregisteredusers
- cnt:countoftotalrentalbikesincludingbothcasualandregistered

Source: Fanaee-T, Hadi, and Gama, Joao, "Event labeling combining ensemble detectors and background knowledge", Progress in Artificial Intelligence (2013): pp. 1–15, SpringerBerlinHeidelberg, doi:10.1007/sl37–40813-0040-3.

6 Data Analytics for Water Resource Engineering

Shivaji Govind Patil and Sivakumar V.

6.1 INTRODUCTION TO WATER RESOURCE ENGINEERING

Water is a cyclic resource with abundant supplies on the earth. Water is a circulating resource that has many things in the world. About 71% of the Earth's surface is covered by water; however, fresh water makes up only 3% of the total water. Indeed, a very small percentage of fresh water is available for human consumption. Water resource engineering is to design and manage the systems based on the water, which includes planning and management of water distribution systems that convey water to the end water users. Also, it is necessary to plan and design the collection systems that transport the stormwater as well as wastewater, managing the surfacewater and groundwater resources, metering the flows in rivers and streams to quantify, model, and analze major water resources projects (e.g., canals, reservoirs, and hydroelectric works) for proper planning, designing, and many other water-related engineering functions (e.g., Zhang and Tang, 2009).

Water resource engineering can broadly be categorized into the three types of (i) groundwater, (ii) hydrology, and (iii) hydraulics. Out of these categories, the groundwater engineering, deals mainly with modeling and utilizing groundwater and planning and framing of groundwater lifting systems. Hydrology is primarily related to the physical boundaries of the watershed and river modeling and understanding interactions of atmospheric, surface, and subsurface water with each other. Hydraulics mainly focuses on the water flow mechanics, pressurized flow, and open channel flow and flow-structure interactions.

6.1.1 ROLE OF DATA ANALYTICS IN WATER RESOURCE ENGINEERING

Water resource engineering encompasses the study of hydraulics, hydrology, environment, and some disciplines associated with geology. Predicting and estimating water resources parameters like sediment discharge, water discharge, rainfall, and runoff and water quality are highly challenging for engineering. It is established fact that water is a significant and precious natural resource and its demand is always increasing day by day and water shortages are now observed as most frequent (Kolhe et al. 2019). On the other hand, the latest developments in various fields such as big data, mechanical learning, and artificial intelligence have begun to provide effective opportunities to use water purification methods and provide the highest quality water supply. Actually, big data concerns almost all the data

DOI: 10.1201/9781003316671-6

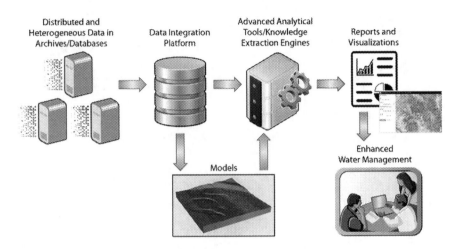

Distributed and
Heterogeneous Data in
Archives/Databases

Data Integration
Platform

Advanced Analytical
Tools/Knowledge
Extraction Engines

Reports and
Visualizations

Models

Enhanced
Water Management

FIGURE 6.1   Data Analytics for Water Resource Engineering.

possessed and then transforms these into knowledge that may be directly employed to manage water treatment and supply activities in a more affordable and easier way. The proper data, correct analytics, and accurate decision framework may pilot water utilities to a higher efficiency. Possessing highly voluminous data without enough comprehensiveness or readiness for its application, it becomes necessary to fine-tune data collection and utilize it into an integrated system of data handling, may prove to be prudent, enterprising, and even to make better decisions. Otherwise, employing big data in water treatment and supply systems may remain only at the primary stages. Considering future trends, pooling data, and using different appropriate analytical tools to make predictions where the analytical should head to become more proactive may prove a great leap towards the advancement of field of water treatment and supply. Figure 6.1 shows the data analytics process for water resource engineering.

Water emergence, and energy and food, as one of the three major interconnected nad global environmental issues present many challenges and opportunities for the use of data on problems such as distribution of water supply and construction of water networks, modeling and predicting floods, rivers, urban and coastal waters, and sanitation. Similarly, advances in GIS, remote-sensing techniques, and weather forecasting techniques mean that environmental data are becoming more and more simultaneous as the demands for solutions and tools to work on these problems become more urgent.

6.1.2   SUSTAINABLE WATER RESOURCE ENGINEERING

Several problems are facing the water-related world: floods, droughts, access to drinking water, access to agricultural production water, pollution of water resources, and lack of basic sanitation. The public and private sectors, both profitable and non-profit, all work tirelessly to develop tools to improve decision-making processes to better predict, monitor, control, and prevent the problems outlined above. Advances in data science, mechanical learning, and artificial intelligence, after access to large and complete data sets, enable the implementation of new methods that support traditional water conservation and sustainability efforts. In short, data science in water has the potential to bridge the gap between surveillance and physically based predictive models. Proper management of water resources is an important task that must be undertaken taking into account various time scales. In particular, the reorganization of water resources is planned periodically to maximize water efficiency and to conserve field strength in active agricultural lands and to stabilize water systems. Many researchers consider irrigation to be one of the key factors in improving agricultural sustainability and depleting groundwater. In these volatile and demanded markets, groundwater helps to strengthen the consolidation of agricultural production globally (Amaranto et al. 2019).

6.1.3 TYPES OF DATA ANALYTICS IN WATER RESOURCE ENGINEERING PERSPECTIVE

Data analytics is a broad field of learning. The methods such as descriptive, diagnostic, predictive, and perspective analytics are used in many applications including water resource engineering. Each method has a different goal and a different location in the data analysis process (Figure 6.2). These are also key data analytics applications in the application of science to business analysis. There are many ways to make sense out of data. The choice of method depends on the queries and the details they want to get from the database.

A major role of data analytics is to understand historical observations or events (descriptive analytics), what will happen based on historical predictions (predictive analytics), and which is the best solution for uncertainty (descriptive analytics). Interpreting these approaches to groundwater research will lead to an understanding of the basic relationships with the various hydrogeological development processes based on available information (descriptive analytics), using this data to predict groundwater conditions (predictive analytics) and to understand which actions are

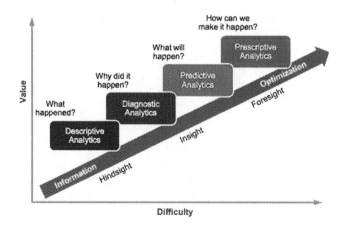

FIGURE 6.2   Data Analytics Maturity Model.

moving forward (Figure 6.2). Therefore, data-driven solutions are more important where analysis will be based on data trends rather than theory (Gaffoor et al. 2020).

Data analytics is a very important part of the value chain, which incorporates data into driving data acquisition processes, as we move from raw data to useful information (Gaffoor et al. 2020). Many of the analytical methods are discussed in subsequent paragraphs. Emerging data-driven modeling using artificial intelligence and machine learning are perhaps the current trend in terms of extracting useful information from the data. Analytical models are highly valuable to the groundwater management studies. For example, descriptive and predictable models may allow for the simulation of current and future conditions in groundwater conditions, while prescribed models may allow for clarification of the impact of various management decisions. Finally, the information used should be distributed in the form of maps, statistics, and tables, etc. This information can be used literally or incorporated into decision support systems, early warning systems, etc. (Gaffoor et al. 2020).

6.1.3.1 Descriptive Analytics

Descriptive analytics serves as an initial encouragement to clear and summarize data analysis methods and techniques. It displays "what we know" as current user data, real-time data, previous analytics data, and big data (Talend, 2021). Descriptive analytics looks at statistical data, which has occurred in the past and helps analysts understand how it works by providing context to help participants interpret data (LA, 2021). Results will be in the form of data visualizations like tables, reports, maps, graphs, charts, and dashboards. For example, a high number of water level monitoring wells are lower than the water level and provide data with corresponding statistics like volume, water level in cubic meter, geology, soil, rainfall, geomorphology, geographic setting, etc.

6.1.3.2 Diagnostic Analytics

Examination receipts for diagnostic data describe an additional step and provide in-depth analysis to answer questions such as why did this happen? Often, diagnostic analysis is called causal analysis (LA, 2021). The diagnostics analytics processes such as data mining, data discovery, drill down, and drill through. In the water level monitoring example, diagnostic analytics would explore the data and make correlations (LA, 2021). For example, it supports identifying which are monitoring wells that have the same level of water table and discharge due to similar hydrogeological settings of the area. Therefore, it shows clarification of the same pattern of groundwater discharge in the area.

6.1.3.3 Predictive Analytics

Predictive analytics and current and historical data are used for predicting the future. The future behaviour and events of variables can be predicted with the predictive analytics models (Kumar and Garg, 2018). Scores are provided by predictive analytical models, where high scores indicate a high probability of event incidence and low scores indicate low probability of event incidence (Kumar and Garg, 2018). Historical and transaction data patterns are used by models to find solutions to problems and to add value to an existing process. These types help to identify risks and opportunities for everyone who makes decisions. With the increasing focus on decision-making solutions, forecasting models have become dominant in the field of water resource engineering (Kumar and Garg, 2018).

Water management is an important issue pursuing many countries due to many factors including geographic, environmental, political factors distress the water in the regions. Therefore, it is important to develop a predictive-analytics framework to study important questions related to water availability such as quantity, quality, pump operation status, etc. (Bejarano et al. 2018), specifically, to answer the following questions (Bejarano et al. 2018).

- Can we accurately predict the performance of a water pump, water quality, and quantity found in data?
- In what ways groups of objects affect the operation of a water pump, its quality and quantity?
- Answering these questions is important for the social and economic growth of communities in these regions. Answering these questions is important for the social and economic growth of communities in these regions. They provide important information that can be used by authorities to identify and effectively deploy scarce resources. The analysis result provides insight into where you can install new pumps depending on water availability and which pumps need immediate repairs, etc.

6.1.3.4   Prescriptive Analytics

Prescriptive analytics is a process that analyzes data and provides a quick guide on how to improve the performance of many predicted results (e.g., TALEND, 2021). In short, targeted analytics take "what we know" (data) to better understand those data to predict potential, and suggest steps towards data removal (Talend, 2021). The analytics use mathematical models and machine learning algorithms to identify opportunities and recommend actions. These types and algorithms can detect patterns in data.

6.1.4   DATA ANALYTICS CHALLENGES IN WATER RESOURCE ENGINEERING

Water and other water providing sources in all areas are day by day facing new challenges in providing a high-quality supply of drinking water along with maintaining the conversation of resources like costs and energy use to a minimum level (DSC, 2021). Several locations of the automated or smart water meter infrastructure has already been instigated to measure the water consumption which provides highly accurate readings. In addition, employing advanced analytics on the collected data can provide an insight for controlling and understanding water network and managing more effectively (DSC, 2021). However, data analytics are not easy and straightforward methods to derive knowledge information as data coming from variety of sources, sensors, different formats, and high volume and velocity. Therefore, data analytics process demands advanced technology like big data technology, AI, high-end computer and hardware technology, etc., though these

FIGURE 6.3   Data analytics Challenges in Water Supply Resources.

technologies are not always available to small-scale research institutes, industries, and institutes.

Besides the software and hardware challenges, data analytics techniques have great advantages to translate data to knowledge, for example, water management agencies and utilities can benefit greatly if they get reliable data on time. The fault detection leads to early indications of abnormal consumption or wastage of water in the network. In addition, historical data sets aids to understand the prior usage patterns that occurred before or not. These types of analytical methods allow learning of the facts like, if it is appropriate for the current seasonal demand and whether it coincides with what neighbouring families, companies, or other clients are consuming (DSC, 2021). Figure 6.3 shows the data analytics and its challenges in water supply resources.

The availability of data in a huge volume and variety is largely benefits to scientific community, however it poses huge challenges to analyze. A large amount of multifaceted data is coming from many satellites; for example, Landsat, MODIS, IRS, Sentinel, EOS mission, etc. The emergence of data sources such as unauthorized aerial vehicles, small satellites, inter-ferometric performance radars, weather monitoring sensors, and other IoT devices present important details that were not previously available in such volumes. These new sources of information will improve our understanding and strengthen our modeling skills. However, data availability is a problem (Shen, 2018). Extracting understandable but hidden information from an unprecedented amount of data or even devising ways to extract that information can be a daunting task. Common methods of retrieving information often require in-depth knowledge of staff and background work because they often require case-by-case evaluation and correction. Advanced data analysis tools to extract hidden data from large volume data (Shen, 2018). Within the water sector, the benefits of big data and its field of data science applications are slowly being linked to create a new amount of information about business assets that control existing assets through increased trust, efficiency, supply chains, and customer relationships.

Monitoring and management of water resources has become a priority for local planning and decision makers at the international level. The complexity of water resource problems is reflected in the interaction of certain physiological conditions (Ridouane et al. 2015; De and Singh, 2021). Problems with water resources include irrigation maintenance, water management, construction of dam storage and mitigation dams, river management, pond management, multi-pond operation, pollution control, etc. With the advent of new technologies such as big data, AI, GIS, remote sensing, social media, crowed sourcing, and IoT have led to significant changes in the amount of data produced daily in real time, this information is of great value, if collected and used effectively (De and Singh, 2021). This chapter discusses advanced analytical, statistical, mathematical, and graphical methods that can be used to convert data into relevant information related to data analytics and water resources (Ridouane et al. 2015). Also, we present our suggestion to conduct data analytics techniques and methods in water resources engineering. In conclusion, the impact of using these techniques in water resource management was discussed.

6.2   ROLE OF BIG DATA IN WATER RESOURCES

Big data is a term used to describe high volume, velocity, and variability. It requires new technologies and techniques for capturing, storing, and analyzing big data. Recently, due to the increasing use of the Internet, digital devices, mobile phones, sensors, social networking sites, satellite images, etc. a large amount of large data (over the petabytes) is produced per minute or second (Agrawal et al. 2012). One can use this data to make decisions, determine trends, forecast the weather, plan good resource management, etc. Big data creates a new generation of data management support. The key to obtaining value from big data is the use of analytics (Agrawal et al. 2012). Information is used to improve decision making, provide understanding and discovery, support and implement processes. Big data has the likely to transform not only research, but also science and technology. This section discusses big data concepts, major data analysis strategies, data sources, data collection, storage, data retrieval, and analytical challenges for various water resource engineering issues including water conservation, water management, dam construction for mitigation and/or conservation purpose, river management, basin management, groundwater rehabilitation, water conservation measures, pollution control, etc. (e.g., Ridouane et al. 2015).

6.2.1   BIG DATA CHARACTERISTICS

Big data is often described as the ability to manage and analyze very large data sets with traditional data processing tools and techniques (e.g., Ohlhorst, 2012; Chen, 2014; Ryan et al. 2021) have gone beyond. Big data is used to improve decision making, provide insight and innovation, and support and optimize

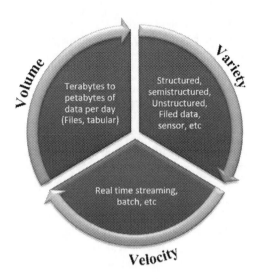

FIGURE 6.4   Big Data Characteristics (Modified after Chandarana and Vijayalakshmi, 2014).

processes (e.g., Anamika et al. 2016). Big data have certain characteristics that set them apart from larger data sets in general. These characteristics are identified as V3, which is high volume, velocity, and diversity (Gandomi and Haider, 2015) (Figure 6.4). Large data volumes contain enormous amounts of data, typically beyond the limit of one terabyte, although these change with time, region, data type, and use case. Large data speeds are generated at an exceptionally high rate, which means that large data volumes increase exponentially over time. Diversity is another characteristic of big data that is composed of different types of different data from different sources.

Science is driven by data. Big data is growing rapidly, with only a 5% increase in global information technology spending currently at an estimated rate of 40% increase in the amount of global data generated per year (Guo et al. 2015). Nearly 90% of the world's digitized big data was captured in the past two years (Bernard, 2015). Revolutionary potential is not only in big data research but also in science and data analysis. Envisage a system with massive database access, where detailed measurements of every event that takes place in the world are collected. Big data is not derived from null; it is recorded from certain data generating sources such as scientific experiments and simulations, which can easily generate petabytes of data per day. Much of this data is of no interest, and can be filtered and compressed by magnitude order. One of the main challenges is to define these filters in such a way that they do not discard useful information (Mayer-Schönberger and Cukier, 2013). Data analytics play a major role in extracting useful information from large data sets, which search for various patterns and other useful information.

Big data analytics is a set of technologies and techniques that require new integration to reveal large hidden values from large data sets that are different, more complex, and larger than normal. It mainly focuses on solving new problems or old problems better and more effectively. A few years ago, big data usage was limited to information technology, and now it has become a developing field in almost all engineering technology. But for water managers/engineers, big data can make a great promise in many water-related projects, such as setting up good water systems; the discovery of changes in the environment through large sensory systems remotely with geospatial systems; predicting and predicting natural and man-made disasters; irrigation planning; reducing pollution; and learning about the effects of climate change (Agrawal et al. 2012; Ryan et al. 2021).

6.2.1.1   Sources of Big Data for Water Resource Engineering Applications

6.2.1.1.1   Geospatial Data

Geographic data is traditionally collected using ground surveys, photogrammetry, and remote sensing. Traditionally, this data has been classified into three formats: raster data, vector data, and graph data (Lee and Kang, 2015). Raster data typically includes land images obtained by unmanned aerial vehicles, security cameras, and satellites. Vector data is generated in large quantities for different types of applications using multiple sources. Graph data is collected and generated from various map sources, including surveys filed discussions with individuals, similar maps, existing survey maps, etc. In recent decades, the rapid development of information technology has led to the rapid development of data from various sensors, leading us into the big data age (Cui and Wang, 2015; Jingdong et al. 2019). With the discovery of new sensors, new methods have been developed to collect geospatial data, which has led to new data sources and local data types. Currently, the availability of data obtained by so-called voluntary location data and location data from geo-sensor networks is increasing exponentially (Xingdong, Deng 2019). Until recently, while dominant data sets dominated the topographic domain, these new data types expanded and enriched geographic data with background diversity and more user-focused.

Geospatial big data, along with geo-location information generated from sensors such as smartphones and handheld global positioning system (GPS) devices, are already rooted in our daily lives and are responsible for disaster response, environmental monitoring, and water resource management (Deng et al. 2019), demonstrate great efficiency in practical applications like urban traffic, water resource engineering, etc. (Jingdong et al. 2019). With the development of new technologies such as smartphones and wireless networks, people can now post and share information on the Internet, anytime, anywhere, such as check-ins, photos, and shopping comments with geographical coordinates. There is face-to-face communication and human activities create a new form of interaction (Xingdong, Denget al. 2019). In the context of the rapid development of social media and big data, these intuitive sharing behaviours often produce geo-tagged big data and are associated with unnoticed phenomena such as human activity dynamics (Deng et al. 2019) and provide potential solutions for identifying and exposing patterns (Jingdong et al. 2019). Some selected sources of geospatial data are given in Table 6.1 and 6.2.

6.2.1.1.2   Internet of Things (IoT)

The Internet of Things (IoT) is a network of physical objects or objects embedded within software, electronics, sensors, and network connections. It enables objects or

TABLE 6.1
Geospatial Data Sources

S. No.	Data Source Name	Description
	ISRO Bhuvan	Indian Geo-platform of ISRO by National Remote Sensing Centre (NRSC). https://bhuvan.nrsc.gov.in/home/index.php
	Open Data Archive Bhuvan	GIS – Open Data Archive from NRSC. https://bhuvan.nrsc.gov.in/home/index.php
	OpenStreetMap	Global crowd sourced data includes points of interest, buildings, streets and street names, boat routes, etc.
	Data Gov	Indian open source data. https://data.gov.in/
	India Water Portal (IWP)	India Water Portal shares information working papers, reports, data, articles, news, events, opportunities and discussions about water. https://www.indiawaterportal.org/datafinder
	ASTER GDEM	30-meter resolution elevation data.
	India-WRIS	Water information in the public domain – India-WRIS Project Initiative for the purpose of disseminating information to the public domain forms the most important elements in the management of water resources. https://indiawris.gov.in/wris/
	HydroSHEDS	Global hydrological data based on SRTM elevation data. River networks, water boundaries, water indicators, and flow collections are included.
	Global Agriculture Lands	NASA's agricultural land use in 2000 from MODIS and SPOT.
	Global Irrigated Area Map (GIAM)	Vector mapping for global rain-fed and rain-fed plants.
	Historic Croplands Dataset, 1700–1992	Documents on historical changes of cropland cover (1700 to 1992).
	Global Reservoir and Dam (GRanD) Database	Reservoirs with a storage capacity > 0.1 cubic km.
	USGS – Earth Explore	The USGS Earth Explorer data centre is your one-stop shop for locating data sets from the USGS wider collections.

objects to first collect and modify data (Table 6.2). Technology is growing rapidly from wireless sensors to nanotechnology and IoT development depends solely on such growing technology. The IoT focuses mainly on how common objects or objects can be heard, smell, see the living world around you, and connect them to share information (Qihui et al. 2014). Items here include cell phones, refrigerators, air conditioners, sensors, actuators, RFID tags, and much more, which use different coping systems, are able to work together, and collaborate with their neighbours to achieve the same goals. The main component of IoT is an embedded system. Various applications of IoT include smart home, smart waste management, smart

TABLE 6.2

Sources of Data in the Water Resource Domain from a Big Data Context (Source: Gaffoor et al. 2020)

Source	Description	Characteristics
Remote sensing	Satellite, airborne or ground-based earth observation	Unstructured and structured Multidimensional Voluminous Regional
Internet of Things	Data available from connected devices	Unstructured and structured Heterogeneous Multidimensional Local
Field activities	Data generated from field activities such as monitoring, drilling and pumping activities	Structured data formatLimited coverage (spatially and temporally) Local
Historical	Legacy reports, maps, and documents	Unstructured Local or regional Text or images
Social media and the web	Data available on webpages and social media post	Textual, images, videos, or audio Multidimensional Heterogeneous Voluminous Local
Computer simulation	Data generated through computer-based models	Unstructured and structured Multidimensional Voluminous Regional

cities, smart water management, smart logistics, smart emergency control, smart agriculture, smart health, and many more. IoT is mainly driven by analytics, artificial intelligence, and data pools. Data pools include sensors, actuators, communication, people, and process. Practically all such above-mentioned IoT applications generate massive data and they are collected from vehicles, cameras, and all such heterogeneous sensors, which provide the data like video, voice, text, and so on. Figure 6.5 shows data sources and flow for smart water monitoring.

6.2.1.1.3 Crowed Source

Apart from the many tools and sensor-based data, crowd source approaches have recently become increasingly effective in researching and using to fill data gaps (Weeser et al. 2019). Information collected by citizens can help to create new information and can support efforts to find human impacts on multiple domains (e.g., water source). More recently, funding crowds is an emerging way to deal with the growing challenges associated with data collection. The crowd source method was often used as a problem-solving model (Brabham, 2008) and the method of outsourcing and

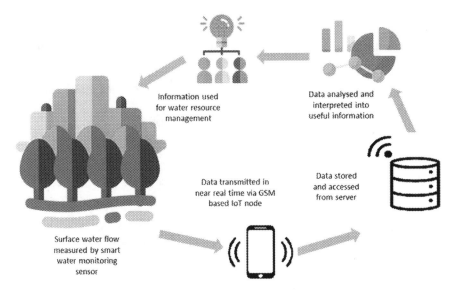

FIGURE 6.5　Smart Water Monitoring Model and Data Flow (modified after Fell et al. 2018).

distribution (Howe, 2006). It is currently regarded as a form of 'citizen science', involving citizens in science, from data collection to basic validation (Zheng et al. 2018) (Figure 6.6). For example, manual reporting of rainfall data to weather services and monitoring of river water level, soil erosion, dam management, irrigation management, crop type, and acreage information, etc. Data statistics play a major role in extracting useful information from a large number of demographic data.

6.2.1.1.4　Field Data

Field data are collected through primary surveys or filed survey (Table 6.2). Primary surveys are conducted at the local level to gather information. Well-planned field survey is an important component of the geographic enquiry that enhances the data

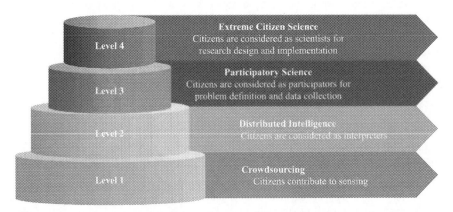

FIGURE 6.6　Levels of Participation and Engagement in Citizen Science Projects.

analyzing capability (NC-NIC, 2021). A field survey not only improves our knowledge, it also shows the spatial pattern, distribution, and relationship at local level. Local-level information is challenging an option through secondary sources; therefore a field survey aids in collecting local-level information at greater level through interviews, measuring, drawing, observation, discussion, etc. Therefore, a primary survey is carried out with predefined objective to get required information for problem solving and decision making. These studies empower the investigator to understand the state and procedures in totality and at the place of their occurrence (NC-NIC, 2021). Data are collected with the help of multiple instruments like GPS, digital camera, handheld sensors, laptop, mobile, etc. It is stored in varying format and methods. Data type will differ based on the purpose of study, for example water resource management to construct checking dams, along with the river soil data, local slope, sediment history/cross section, river pattern, and rock type, etc., to plan effectively. Figure 6.7 shows water-level measurement methods in the field.

6.2.1.1.5   Social Media and the Web

Social media and web content have brought about rapid changes in society, from our communication systems and grievance systems to our electoral and media

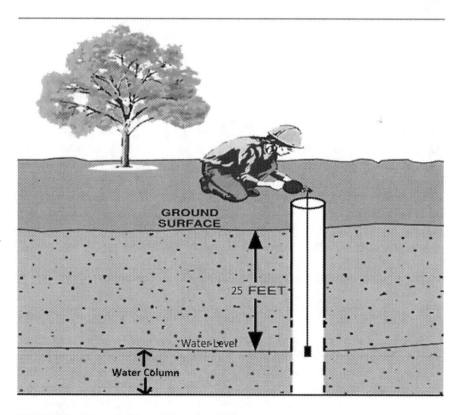

FIGURE 6.7   Example Shows Water-Level Measurement Methods in Field (modified after Schreffler, 1997).

outlets. It is increasingly being used by individuals and organizations in both the public and private sectors. More than 30% of the world's population is on social media; therefore, a large amount of data is available in the science and engineering application. Information is largely available to analytics from many social networks such as Twitter, Facebook, the web, etc. It is an important data source for emergency response such as hydrological science (e.g., flood map) (Ilieva and McPhearson, 2018). Important information from the ground up helps to raise awareness of the state of hydrological events management. Social media data considers citizens as natural sensors, not only protected from external factors such as clouds and fog but also features of real-time space distribution and fine-tuning, which leads to compensation for remote satellite imagery in space and time resolution (Ilieva and McPhearson, 2018; Li et al. 2017). It is proven that satellite limitations can be slightly overcome by communication data by mixing the data sources of the area under investigation. Table 6.1 listed data sources on the water source domain from the main data center.

6.2.1.1.6 Historical

Historical data, in a broader context, collects data about past events and events related to a particular topic or event. By definition, historical data includes most of the data manually or automatically generated by the data collection team. Sources, among many possibilities, include embedded survey, satellite/sensors, media releases, log files, project and product documents, and email and other communications. Other data sources for geospatial history are listed in Table 6.2.

6.2.2 BIG DATA ANALYTICAL TECHNIQUES

Advanced data analysis techniques are used to convert big data into intelligent data extraction of important data. Smart data is the data that is filtered, cleaned, and prepared for context. These data are being analyzed for insights, in turn, to provide improved decision making. Big data utilization is growing in the modern digital world. Annual growth rate for big data and technology services is expected to increase by 36% with global capital gains and business analysis expected to increase by more than 60% (Khan and Ayyoob, 2018). For better decision-making process big data are converted into smart data or small problems, using advanced technology like statistical, AI, ML, data mining, blockchain, etc. The following section discusses in detail about big data analytical techniques are used for deriving smart data.

6.2.2.1 Statistical

A statistical method is a type of conventional data analyzing techniques that is designed to perform the function of regression, categorization, segmentation, clustering, detecting anomaly, prediction, etc. For example, groundwater modeling analyzes multivariant regression analysis (e.g., geostatistics) can be used to measure causal relationships between a series of variables, which can be used to predict the outcome of a dependency variation (Gandomi and Haider, 2015). For monitoring the groundwater system, water quality and flow statistical techniques are widely used; however, this method is not capable of handling highly dimensional, heterogeneous, and noisy data (Gandomi and Haider, 2015). Additionally, it is challenging to implement in parallel-processing/HPC environments, which is essential when trading with big data (Chen and Zhang, 2014). However, scientific community is integrating this method into advanced data analytical discipline in handling structured data.

6.2.2.2 Data Mining

Data mining is an analytical technique for extracting relevant information from large amount of data for the purpose of decision making (Figure 6.8). It is widely used to understanding the pattern, relationship, categorizing the data, etc., (e.g., Ali et al. 2016). Many statistical and ML techniques are used in data mining application. Data mining techniques can be applied on structured and unstructured of data, volume, types like image, video, text, etc. Many data mining tools such as R-Programming Tool, WEKA, RATTLE, JHEPWORK, ORANGE, and KNIME are used including in water resource engineering applications for gaining knowledge and understanding the hidden content from the data. In the water resources domain, data mining techniques are used in rainfall analysis, weather report analysis, forecasting drought, predicting the in/out flow, temperature pattern study, etc. For example, reservoir management is a critical task for engineers in a particular watershed and highly depends on downstream water requirements, inflow, and storage capacity. To ensure the balanced or optimized water supply, historical reservoir operation data are analyzed, and derived the hidden pattern based on reservoir rule curves (e.g., Ahmad and Simonovic, 2000). This model will extract the knowledge in the form of 'rules of pattern or if-then' rules and subsequently used for simulation of reservoir operation followed by the decision-making processes (Mohan and Ramsundram, 2013).

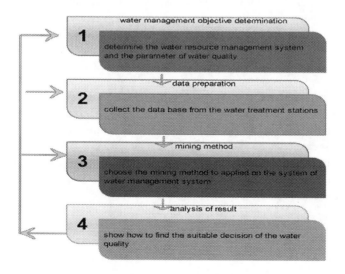

FIGURE 6.8 Processes of Data Mining (adopted from Hussein et al. 2014).

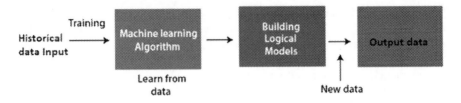

FIGURE 6.9   Simplified Model for ML Work Flow.

6.2.2.3   Machine Learning

Machine learning (ML) is a sub-field of data science that focuses on studying computer algorithms that will advance inevitably through experience and by the use of data. Sample data or training data sets are used to build the models in this learning method for the purpose of predicting and making decision. The ML method experiences supervised and unsupervised learning and reinforcement leaning. It is well suited for modeling using the big data sets and breaks down the records at different levels. ML algorithms improve performance by mechanical retrieval of information from experience. ML aims to provide increasing levels of automation in the engineering knowledge process, eliminating time-consuming human activities with automated techniques that improve accuracy or efficiency by acquiring and exploiting common training data (Simon and Langley, 1995). Considering this, ML has been used for a long time in mathematics, statistics, and engineering with models or algorithms like linear, polynomial regression, time series regression, etc. Figure 6.9 shows a simplified model for ML workflow.

Recently, utilization of ML technique in the water resource engineering field is rapidly growing, for example, for improved reservoir operation of historical data including weather, soil moisture, sediments, erosion, evaporation, and land user and land cover and previous reservoir operational data are collected to gain the knowledge to build the models. Captured knowledge is utilized to train and enhance the model for improved operation of the reservoir in the watershed. Figure 6.10 shows the machine learning process workflow.

6.2.2.4   Visualization

Visualization techniques are playing a major role in effective analysis of the results, modeling, report generation, and decision-making processes. Data can be visualized in many ways such as graphical, chart, GIS map, images, tabular, multi-dimensional (2D, 3D, 4D), etc. Visualization methods allow for intuitive view of data, including geographic content and insights into pattern analysis for expert judgment. Figure 6.11 shows visualization of the water level fluctuation of the area.

The left image shows water level fluctuation in a GIS map and the right image shows water level fluctuation represented in graphical format (Image source: Patil and Lad, 2021).

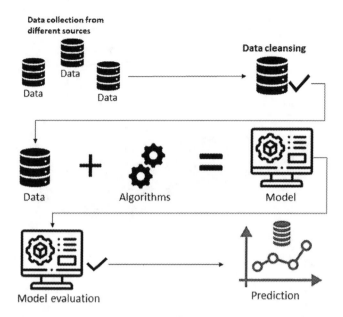

FIGURE 6.10   Machine Learning Process Workflow.

6.2.3   CHALLENGES OF BIG DATA ANALYTICAL TECHNIQUES IN WATER RESOURCE ENGINEERING

Implementing a big data analytical method is a challenging task for domain experts for processing and managing, due its characteristics like a large volume, variety and velocity; therefore, the amount of data, especially machine-generated data, explodes, as quickly as that data grows every year, with new data sources emerging. According to the IMB survey, the world data volume was 800,000 petabytes (PB) during the year 2000 and it is expected to increase beyond 175 zettabytes (ZB) by 2025 (Reinsel et al. 2018).

   Data collected from various methods and sensors like water level monitoring, meteorically, water quality monitoring, spring water monitoring, glacial lake water monitoring, manual reading at filed level, crowd sourced data, etc., are mostly semi-structured to unstructured. These data are not uniform as various IoT devices and methods are used; therefore, effective modeling is challenging, until data is cleaned in a structured manner. For example, one of the major challenges in water resource engineering is that for a natural disaster like flood prediction and monitoring process, data are gathered in real time and analysis is required to produce results in real time for decision making. Big data handling technology is always not available to application builders. However, recently big data analytical infrastructures like cloud computing, parallel processing, and blockchain in AI are emerging globally.

6.3   ADVANCED COMPUTATIONAL INTELLIGENCE TECHNIQUES IN WATER RESOURCE MANAGEMENT

This section discusses how CI methods, models, and algorithms are used to manage water resources that can accelerate the allocation of water resources in science and

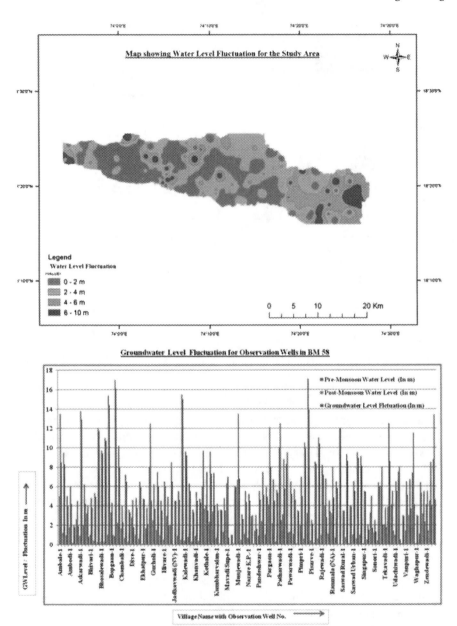

FIGURE 6.11 Visualization Shows Water Level Fluctuation of the Area (left and right image).

the use of water resources for meaningful development (Zhengfa et al. 2010). For example, clean water is one of the foundations of our lives, so we need to always be successful in getting better and better quality water. Data-driven water management now integrates with artificial intelligence to fundamentally change the approach to ensuring clean water.

6.3.1 Artificial Intelligence Techniques

The purpose of this section is to provide a framework for the artificial intelligence (AI) field. According to AI's father, John McCarthy, it is "the science and engineering of intelligent machines, especially intelligent computer programs". AI work is largely empirical and engineering orientation. The AI system is growing computational techniques, which are developed, undergo experimentation, and are refined for wide applicability. Subsequent sections will discuss how artificial intelligence techniques are used in water resource management. The section will illustrate the use of AI in several small examples but omit the detailed case studies of large applications. An abstract understanding of the basic idea should facilitate understanding AI system to open up the AI concepts in water resource engineering.

6.3.1.1 Artificial Intelligence in Water Supply (Example 1)

Smart data-driven requests have become a hindrance to daily life. Water resources can benefit from the innovation of digital technologies to improve their performance (Jenny et al. 2020). By connecting the power of data analytics through artificial intelligence algorithms, water utilities can enhance the service delivery more economically. The water services industry and the company face challenges in dealing with scarce water, which is one of the key operational parameters in determining the performance of a water resource in reducing both physical losses (e.g., for example, water leak detection techniques) are based on the combination of special equipment (acustic sensors, gas tracers, etc.) and human skills. The current practice is to install AI on one of these hardware (for example, acoustic correlators) to replace people with data translation (water leaks).

The AI-based approach has many advantages over the traditional method, as it removes the difficulty of connecting in real time to digital water systems and tests the performance of a water distribution network with a larger data set than can be done in real time, taking into account the pilot's average time in the implementation phase. With the advancement of hydraulic modeling of the water distribution network, it is now possible to detect leaky pipeline segments by numerical methods, provided that hydraulic models are provided with a sufficient amount of field data measured such as pressure, flow, and use of nodes (ADB, 2020). All the variables of water heating in a water distribution network reflect a certain type of integration, made up of the basic rules of compression volume (fed power and size, as defined in Bernoulli's policy), which represent the basis of any water distribution system. Numerical algorithms designed for water leakage detection are intended to detect specific local and temporary patterns as well as irregularities in the flow and pressure values in specific areas of the water distribution network. This is to extract information on weight loss and trade (ADB, 2020). Calculation methods need to be measured with field data; when the supervisory control and data acquisition (SCADA) monitors the water supply system, numerical methods become less expensive investments (see Figure 6.12).

6.3.2 Machine Learning

ML is an important part of the growing field of data science that automatically enhances data experience and usage. ML algorithms create a model based on

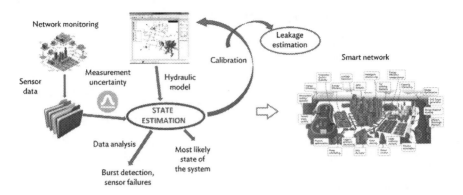

FIGURE 6.12 Network Analysis System.

sample data, known as training data, in order to make predictions or decisions without explicitly planning to do so. Machine learning algorithms are used in a variety of applications (Yi et al. 2015). Using mathematical methods, algorithms are trained to perform categories or predict, to obtain important data within data mining projects. This information also furthers the decision making within applications and businesses, which has a positive impact on key growth metrics. As big data continues to grow and grow, the market demand for data scientists will grow, which requires them to help find relevant business questions and later with data to answer those (Yi et al. 2015).

6.3.2.1   Water Quality Management (Example 1)

Water quality management is important for all living beings. In the last few decades, quality of surface and sub-surface water has severely been affected due to various source of pollutant like industry, human and agricultural waste, etc. Due to a high concentration of chemical and harmful organisms in the water, this leads to unsafe for human consumption and affect the ecosystem (Adusei et al. 2021). Monitoring, managing, and predicting water quality through many techniques are used globally; however, the ML model plays a significant role in gaining accuracy of results. For example, to identify the water quality parameters, water samples are collected at specific locations and analyzed in the lab; however, this method is very time consuming and tedious task in regional scale. Recently satellite remote sensing provides the global coverage in good spatial, radiometric, temporal, and spectral resolution therefore, it is easy to identify the water quality parameters and developing predictive models using ML techniques. ML models like SVM, RF, and NN have been used by various researchers (Sagan et al. 2020; Prasad et al. 2020; Naghibi et al. 2016) to obtain better accuracy as ML techniques have the ability to ability to learn automatically from data and identify hidden patterns from satellite imagery. The main purpose of utilizing these methods by the water resource engineers is that it has the capability to identify pixel level correlation and estimate the spatial distribution in the specific area. Figure 6.13 show the spatial distribution of water quality parameter that was derived from satellite imagery based on ML methods.

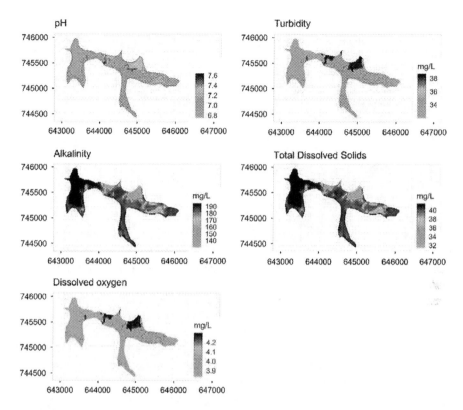

FIGURE 6.13   Water Quality Parameters Identified Using ML Methods from Reflectance Satellite Imagery (Adusei et al. 2021).

6.3.2.2   Reservoir Management (Example 2)

Hydrology and water resources systems are non-stationary due to their characteristics of time series. Therefore, it is necessary to use methods that can model the non-stationary behaviors of environmental variables to optimize water systems (Milly et al. 2008). Linear parametric approach leads to poor performance results if it is tested with unseen data sets; therefore, it is not suitable for long-term forecasting. Studies proved that machine learning models are more suitable for learning the non-linear dynamics and non-stationary behavior of water resource systems (Pulido-Calvo et al. 2003 and Nourani et al. 2009).

For example, Ticlavilca and McKee (2011) used an important multivariate variety machine to develop multiple future forecasts of daily releases from multiple Sevier River Basin dams in Utah. Their model predicts the drainage of two ponds simultaneously with previous details of the history of drainage, diversions in the lower canals, climate, and flow of streams. Their results have shown the effectiveness and robustness of the machine learning method for predicting the release of multiple pools. They also proposed an ML model for irrigation water supply in the Sevier River Basin in Utah. They introduced a strong ML approach to predicting short-term deviation needs of the three irrigation canals. Their model saw patterns between

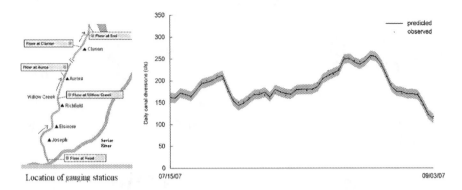

FIGURE 6.14 ML Model-Based Output Versus Observed Results for Reservoir Operation (Ticlavilca and McKee, 2011).

multivariate discharge (future diversion irrigation requirements) and multivariate installation (previous data on irrigation need and weather information). A major problem with water supply to the vessel is the response of an inefficient operator to the much-needed temporary change due to the time remaining between the discharge of the dam flow and its arrival in the diversion irrigation canals (Ticlavilca and McKee, 2011). Therefore, the model predicts that short-term diversion requirements may have the potential to assist pool and trench operators in making real-time decisions and decisions for the management of water resources available in the borehole. Figure 6.14. show the ML model-based output versus observed results.

6.3.3 DEEP LEARNING (DL)

Deep learning considered a subfield of ML, which is based on learning on its own by examining computer algorithms (Figure 6.15). DL works with artificial neural networks that designed to operate similar to humans think and learn. DL learning

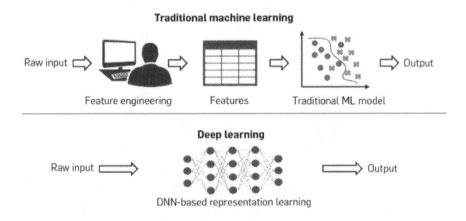

FIGURE 6.15 Shows DL Method.

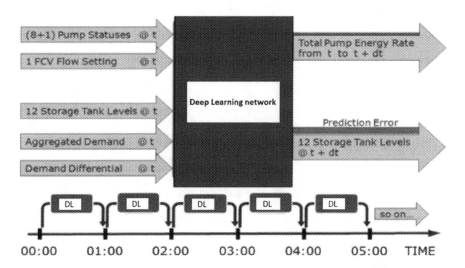

FIGURE 6.16 Deep Learning Model for Extended Period Simulation of a Water Distribution System (modified after Zheng et al. 2015).

provides a black box method for learning from complex and high-quality data to incorporate solid and terrifying insights and thereafter reduce human intervention.

6.3.3.1 Water Distribution Systems (Example 1)

Providing sufficient quantity and quality of water is a key challenge for public water distribution bodies. The advancement of DL methods will offer the effective management of water distribution that leads to minimize the total expenditure. A smart water network emerges as an integrated way to enable the city to improve the efficiency of its water distribution system. This network enables continuously monitoring and diagnosing remotely that provides water consumption data to the customers in regular basis. For example, Zheng et al. (2015) modeled and implemented the smart network using a deep learning framework for extracting intelligence from the data in an efficient manner. This information is used for simulating and predicting for building optimal smart water network and management (Figure 6.16).

6.3.4 Fuzzy Logic

Fuzzy set theory, suggested by Zadeh (1965), is a compilation of classical theory. The logical illegitimate representations found in the fuzzy set theory attempt to take the form of human representation and discuss real-world knowledge when confronted with uncertainty (Zadeh, 1965). A fuzzy set can be mathematically defined by giving each potential person in the entire speech space, the amount that represents its category of membership in the incorrect set (Singh et al. 2010). Figure 6.17 shows the general fuzzy logic architecture.

Fuzzy logic assigns membership values to locations that range/grade from 0 to 1. Fuzzy logic has similarities to Boolean logic. Boolean logic results are restricted to 0 and 1, whereas fuzzy logic returns values between 0 and 1 (Figure 6.18).

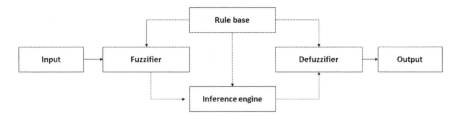

FIGURE 6.17 Fuzzy Logic Architecture.

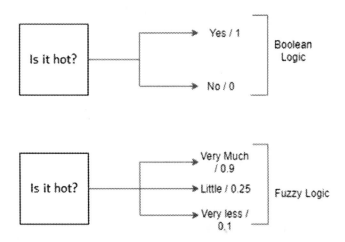

FIGURE 6.18 Shows the Example of Fuzzy and Boolean Logic (modified after GG, 2019).

Therefore, fuzzy logic is similar to human thinking and instead of just true or false; it defines the value based on degrees of truth. Fuzzy membership function assigns the fuzzy membership values are most important for analysis of influence of various thematic layers.

6.3.4.1 Groundwater Mapping (Example 1)

The fuzzy logic method is used to delineate the groundwater potential mapping research and application. The prospects for groundwater potential zone in the selected study area will be explored through the integration of thematic layers/maps in fuzzy overlay function. In order to produce the potential groundwater zone map, the interrelationship of thematic layers and their influence on groundwater system will be analyzed and according to their weightage the fuzzy membership values have been assigned to various classes in a thematic layer. Fuzzy membership function curve supports defining how each point in the input interval can be mapped to a membership value from 0 and 1. This function helps to determine the influence of a parameter between lowest to highest. Many researchers carried out the study on delineating the groundwater potential mapping zone using fuzzy logic methods (Pareta, and Baviskar, 2020; Rajasekhar et al. 2019; Mohamed and Elmahdy, 2017; Bhowmick and Sivakumar, 2014). Figure 6.19 shows fuzzy models and methods for

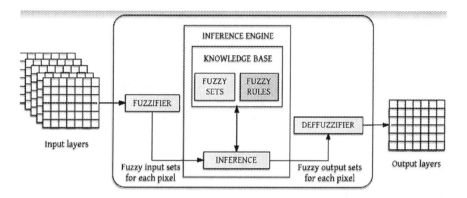

FIGURE 6.19 Structure of the Fuzzy Overlay Method for Groundwater Potential Zone Mapping Using Geospatial Technology (Bhowmick and Sivakumar, 2014).

groundwater potential zone mapping using geospatial technology. Figure 6.20 show the groundwater potential zone map derived based on fuzzy logic approach.

6.4 PREDICTIVE MODELS

Today, planning, designing, and managing water resource systems certainly involve the predicting impacts. Model output is based on structure, input data, and other

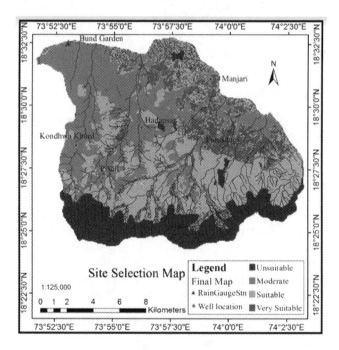

FIGURE 6.20 Groundwater Potential Zone Map Derived Based on Fuzzy Logic Approach (Bhowmick and Sivakumar, 2014).

assumptions that indicate the real-world system with different scenarios. A predictive model result is based on assumptions of inputs that will have big impact on a predicted future event. Therefore, any quantitative study is considered/taken as a part of information which will be used for overall planning, management, and decision-making process. A model result will be used to enhance our understanding about the system and applications. This section discusses how predictive models are used for water resource planning and management, type of models, merits and demerits, applications, etc.

6.4.1 Water Quality (Example 1)

Water quality monitoring, management, and predicting the future scenario is urgent need as it is prime important for the ecosystem. Water contamination due to climate, human activities, and non-treated water discharge are adding more stress on quality of water (e.g., De and Singh, 2021). Globally, many models are designed for water quality management, such as QUAL2E, water quality analysis simulation, and U.S. Army Corps of Engineers' Hydrologic Engineering Center-5Q, etc. Recently, the use of data-driven models, such as the adaptive neural-based fuzzy inference system (ANFIS), neural implant networks (ANNs), and the gene expression program (GEP) has become a different approach in many studies (e.g., Verma et al. 2013). Artificial intelligence (AI) has been used in many water-related studies for example, water quality modeling and water management applications (e.g., Zaman et al. 2014; Nemati et al. 2015; Khudair, 2015; Kanda et al. 2016; Salari et al. 2018). Many traditional forms are used to mimic water quality by scientific communities. However, the benefits of adopting data-driven modeling techniques over others come from their self-study ability from data thus reducing error.

6.4.2 Hydrologic Prediction and Forecasting

Hydrological prediction and forecasting are emerging area of research for hydrological risk analysis, preparedness, and mitigation. Recently, geospatial technology provides real-time information at global level and advanced computational techniques facilitate data analysis and dissemination near real time. New advances in climate research and modeling have enabled annual weather forecasts with the right accuracy and increased resolution. Emerging technology enables us to obtain accurate hydrological information so it is easy to predict and predict catastrophic events such as surface water resources, floods, erosion, dam collapse, etc.

6.4.2.1 Groundwater (Example 1)

Groundwater is often one of the major sources of water supply for domestic, urban, agricultural, and industrial purposes, especially in arid and semi-arid areas. In order to protect water in the future, sustainable management of groundwater resources in conjunction with surface water has become an urgent need for an hour. Accurate and reliable forecasting of groundwater levels is crucial in achieving this goal, especially in wetlands in arid and semi-arid regions that are easily affected by extreme water events (Huang and Tian, 2015). Water-level measurements from observation sources are a major source of information about the hydrologic pressures operating in aquifers and how these pressures affect water regeneration, conservation, and drainage (Huang and Tian, 2015). On average, groundwater levels are followed by process-based models, based on in-depth knowledge of the recognized system power. They require many additional spatial details about the geological and hydrological structures of the aquifer. On the other hand, in data-driven modeling studies, try to identify a precise map between inputs and system outputs without access to the internal structure of the visual process (Klemen et al, 2018). After gaining significant advances in water and water use, such as rain modeling and water quality forecasting, data-driven models are now widely used to solve problems in the groundwater environment (Solomatine and Ostfeld, 2008). Examples of the most common methods used for modeling run by groundwater data include artificial neural networks, ML, vector support machines, genetic programming (GP), and non-regulatory systems such as the M5 model tree, etc. The purpose of the data modeling driven by groundwater level prediction is very accurate, which can be used to help water managers, engineers, and stakeholders to manage groundwater efficiently and sustainably (Adamowski and Chan, 2011). For example, the machine learning strategy model is based solely on data and some domain-specific information is incorporated into the system through appropriate data modification (within the brand-new engineering). The goal in such a situation would be to predict groundwater levels based on the inclusion of temporary data (groundwater history and surface water data, weather data and forecast, land use, groundwater extraction, and other anthropogenic data) and outcomes (groundwater level). The model captures the basic data-based processes without additional user input from the specialist (Klemen et al. 2018). Figure 6.21 shows the flow of the low-level water system based on the model.

6.4.2.2   Drought Risk Assessment (Example 2)

Hydrological drought occurs in a particular geographic area due to lack of runoff. Hydrological droughts are assessed using various indices like rainfall, temperature, evaporation, sub-surface geology, land use pattern, etc. Globally, many models developed for drought forecasting and monitoring meteorological parameters that causes the hydrological drought. For example, the climate prediction center (NOAA – Center for Weather and Climate Prediction) provides a wide range of global weather forecasting data using the climate forecast system, which includes the medium range forecast (MRF) and modular ocean model version 3 (MOM3), from 2002. The drought forecasting models the mid- and long-time predicted rainfall and climatic data area used as an input to understand and derive a drought severity index (e.g., Yoon et al. 2012).

6.4.2.3   Sediment Transport (Example 3)

Generally, sand, silt, and clay sedimentary particles are presented in the watershed. These sediments are transported through streams and rivers from upland to downstream (Adnan et al. 2019). In time, under some conditions in the hydrological system, the sediments are deposited at the bottom of the river due to reducing flow velocity and momentum. The sediments are also deposited in the

1. Data processing
- Inputs: Monthly Precipitation (P), Temperature (T),
Streamflow (S), Irrigation Demand (I), ENSO, PDO, NAO
- Output: Groundwater-level change (ΔGW)

2. Best inputs selection
- Decompose the time series using SSA to extract significant RCs
- Select best RC of inputs by mutual information and genetic
algorithm
- Determine lag time series at maximum correlation

3. Model development
- Selection of neural network architecture, model parameters,
activation function and hidden neurons

4. Training, cross-validation and testing
- Determine optimal weights and number of hidden neurons in
training and cross-validation to avoid overfitting (70% data)
- Test on unused data (30% of data)

5. Model prediction performance
- Evaluate model performance (R, MSE)
- Assess uncertainty in prediction using Monte Carlo method

6. Sensitivity analysis of model parameters
- Calculate relative importance of model inputs by connection
weight approach

FIGURE 6.21   Shows Methodology Flow of Groundwater Level Predictive Model (Sahoo et al. 2017).

reservoirs' quiescent condition by the action of gravity. Therefore, these type of fills/deposits lead to reducing the storage capacity of reservoirs; and block the river/stream free flow the cases of flooding in the watershed. Therefore, it is of important to develop and implement an accurate prediction model for sediment transport. Many forms of speculation on the levels of columns have been improved over time (e.g., Moriarty et al. 2014; Xue, 2012; Blaas et al. 2007). However, the accuracy of the predictions of these species is often uncertain. The mode of transport is complex, and transport models for limited water resources are based on simplifying the assumptions that often lead to major forecasting errors. Data-driven modeling can assist in modeling programs for which sufficient physics information is limited (e.g., Bhattacharya et al. 2005).

6.5 APPLICATIONS OF DATA ANALYTICS IN WATER RESOURCE ENGINEERING

This section focuses on the application of data analytics in water resource engineering include the designing of various hydraulic structures, dams, and also management of the water streams and water ways, flood prevention/protection, environmental management, watershed management, groundwater exploration and development, drought assessment, etc.

6.5.1 MODELING FOR FLOOD MANAGEMENT

Water resource management tools include simulation, optimization, and multi-criteria/multi-objective analysis. Designing simulation modeling tools for flood management in the decision-making process could satisfy multiple goals including multi-objective decisions. Computerized models are needed to be used for longer period to support water-related decision making and water resource management. However, there is a fear that model development is limited to the persons having expertise and who propose such models, which results in the model often remained silent to simplify the dynamics of various interactions viz. economic, social, and environmental and further decision makers lead to defective policy making and also result in taking wrong management decisions. With one of the different approaches, these models get changed with dynamic environments with all of their complexity and also other dependencies like societal. This approach has been adopted to develop independent active robots successfully so that they interact with changing situations and also learn appropriate interactions without constructing environment models. It may be able to incorporate various elements like economic, institutional, socio-cultural, and political elements that have an influence on decision making, addressing non-dominated solutions (e.g., Janusz, 2011).

The motivated learning may provide a useful technique to support planning, decision making, and management of water resources, as it develops a natural way of balancing between various competing needs. Unlike other machine learning methods, it does not follow a prescribed algorithm to optimize its decision; it does not require the environmental model but it is capable of building one and it dynamically adjusts its own actions commensurate with the changes in the environment and its own perception of needs. For example, the need for the water reservoir was an abstract pain developed through the action of irrigating fields to remove the primitive pain signal related to drought. An abstract motivation resulting from the need for water reservoir could be to earn money (to pay for the reservoir), while building tourist attractions may be a way to earn money, and finally building a new reservoir may be motivated by the need to develop tourist attractions. It is necessary that a learning network shall be able to find this out and avoid using such circular solutions. Figure 6.22 shows methodology used for modeling flood risk assessment.

The machine learning method shows how system development stimulates learning of new concepts and at the same time benefits from this learning. It is expected that this natural learning will lead to more accurate models for policies and

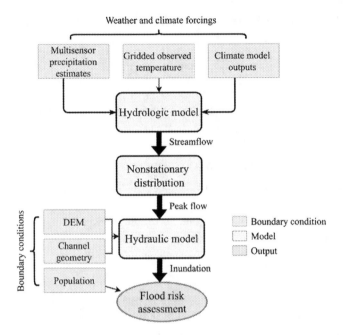

FIGURE 6.22   Shows Methodology Used for Modeling Flood Risk Assessment (Sharma et al. 2021).

actions related to water and through active interaction with human experts will use the provided input data for identifying the influence factors that mostly contribute to water supply, water use, and water contamination or policy-making decisions. A flood is a big flow of stream water, especially a body of water, rising, swelling, and overflowing and not usually an inundation. A flood problem is tremendous and becomes more complex with the passage of time. Damages caused by the flood can be lowered by the proper and timely predicting of floods and proper and timely measures taken to safeguard the structures and lives with effective flood management. Most of the flood studies are made for flood controlling.

The following three types of data are required for developing a model for flood management along with other collateral data and ground truthing.

- Water discharge data on daily basis
- Water level data on daily basis
- Danger water level data
  - i. Daily Water Discharge Data: The mean daily discharge data during a water year is necessary. These data contain the yearly maximum and minimum values of water discharge.
  - ii. Daily Water Level Data: On stuff gauges, the water level of the river is measured five times a day. The mean of the five measurements is considered a mean daily water level. The data also contains the yearly maximum water level data during the monsoon.

iii. Danger Water Level: Danger level data for designated stations is necessary.

After collecting above data, it is necessary to work out:

a. Travel Time from Base Station to Forecasting Station: The travel time for water from the base station to the forecasting station is necessarily to be calculated.
b. Correlations between N^{th} Hour Stage of Base Station and $(N+T)^{th}$ Hour Stage of Forecasting Station.

6.5.2 REMOTE SENSING AND GIS FOR IDENTIFICATION OF GROUNDWATER RECHARGE

In recent times, remote sensing and GIS have proved powerful techniques for identifying the watershed characterization, assessing watershed processes, conservation planning, and management. Various ground features such as geology, geomorphic features, and their hydraulic characteristics may give useful indications of the presence of groundwater. Geomorphology, climatic conditions, and others are the controlling factors of groundwater existence and flow in the geological formations containing hard rocks. These features cannot be seen on the surface with bare eyes but have been picked up through satellite remote sensing with reasonable accuracy.

Recent advancements in remote sensing have opened new avenues for distributed spatial data. With the advantages of availability of spatial and temporal data with remote sensing for large and remote areas during a period provided a convenient technique in assessing, observing, and safeguarding precious groundwater resources. It is a handy tool in the acquisition of spatially distributed data for modeling for the water resources. On the other hand, GIS techniques facilitate integration of multidisciplinary data and conducting the multicriteria analysis with large volumes thereof, within the same geo-referencing scheme.

Remote sensing and GIS provide a wide and very useful platform for analyzing diverse data sets to get convergent outputs, which can be used in decision making in groundwater resource planning and proper utilization. After the invention of super-fast and power-packed computers, ultra-modern techniques for water management have evolved, which are of great significance. Groundwater regime is considered a non-stationery system in which the groundwater is recharged at the surface of the earth and penetrates back to the sub-surface through geological formations. In the groundwater recharge process, various factors like relief, slope, ruggedness, rainfall, irrigation, etc. influences the recharge process besides the geological setup. As such, the framework in which groundwater exists is as changeable as that of the rock formations, as intricate as their structural deformation and geometric history. The lithologic, structural, geometric, hydrologic, and meteorological parameters are more complex and sometimes so obscured that even in the field, investigations are not readily visible. The GIS and remote-sensing technique have emerged as an indispensable tool in groundwater exploration, development, and management.

The remote sensing and GIS can be used in various ways in groundwater exploration, like preparation of regional and detailed maps, selection of sites, estimation of draft, and demarcation of over developed areas. Planning for a combined use of surface and groundwater, selection of sites for artificial recharge also can be done with the help of this technique. However, it is essential to caution that though remote sensing and GIS are extremely useful in exploration of groundwater regime, it is not an ultimate tool and, therefore, field observations are a must for referencing, basic data collection, confirmation, and final conclusions.

Groundwater recharge has become a key word in the era of increasing water demand for numerous uses and therefore it is the need of the time to identify the groundwater storage beneath the ground and estimate them for bringing into use. However, at the same time, there is a need to keep the balance of the groundwater quantity for the future; otherwise, there may be a big issue of water scarcity. It is, therefore, important to identify the groundwater recharge and plan for the optimal use of available groundwater and simultaneously work out the plan for artificial recharge to keep the balance in the groundwater regime. Groundwater recharge can be identified by various methods and the use of remote sensing and GIS become popular due to their multiple aspects of utility in a short time and less cost with more accuracy. The remote sensing and GIS offer a unique capability of integration of spatial and non-spatial data to conduct multicriteria-based spatial analysis and is also capable of handling vast volumes of data sets in less time and at a lower cost in comparison with traditional methods. Using remote sensing and GIS, groundwater recharge distribution can be identified in the form of a groundwater recharge zones map, based on the study of the spatial and temporal distribution of groundwater recharge in the study area, along with field observations for validation of the analysis. All the thematic maps, in raster format, are superimposed by the weighted overlay method using the spatial analyst extension in the tool box of the ArcGIS software. This work involves a multicriteria analysis, i.e., weighted overlay analysis, using ranks assigned to each class of theme and weightages assigned to each thematic layer and their integration in the GIS frame. The final map showing groundwater recharge zones can be derived from this multicriterion overlay analysis, which incorporated all thematic layers, along with their relative class ranks and theme weights.

Assigning weightage to each factor and ranking to each class is the skilful and important part in GIS analysis, which integrates various inputs and delivers output, as the output is mostly dependent on the assignment of proper weightage and ranking. Presently, no standard scale for weighted overlay method is available. The study with overlay analysis involves a multicriterion analysis, wherein investigation is to be carried out with multi-faceted things for determining specific theme, with the help of assigning a rank to the respective class and assigning weightage to the respective feature, depending on the weightage of the theme of the objective. Considering the spatial and temporal distribution of groundwater recharge various themes are to be considered for spatial analysis to generate the groundwater recharge zones map. This map will be useful to suggest artificial groundwater structures suiting the site requirements for enhancement of groundwater recharge structures, at appropriate locations in the region.

6.5.3 MITIGATION MEASURES THROUGH WATER CONSERVATION WORKS

Water is the potion of human life and the water demand is increasing due to the tremendous increase in development activities, along with the increasing population. The principal source for surface water and groundwater is precipitation. The uncertain nature of rainfall and large-scale exploitation of surface water resources has turned the onus obviously on the groundwater. Thus, sustainable development and management of groundwater resources become a challenging task and requires an effective mitigation plan and measures through water conservation works. Major mitigation measures for enhancing the groundwater recharge are directly associated with the artificial recharge structures. Various government organizations, non-government organizations, and other voluntary entities along with local community participation are carrying out various water conservation works suiting the local hydrogeological and other conditions by the way of constructing different types of artificial recharge structures (Kaliraj et al. 2015). Attempting for the artificial groundwater recharge is very much essential as it not only helps to overcome the adverse conditions like scarcity of water for many purposes, due to insufficient or poor rainfall causing the drought-like situation but also helps to add the more water quantity available for the prospective users. Thus, mitigation measures are very essential for improving availability for different water users for keeping the pace of development activities upwards.

Artificial recharge structures can be suggested by taking into the consideration of available non-committed surplus monsoon runoff, to ensure filling of these structures during the rainy season. To achieve the balance between demand and supply, it becomes obvious to go for mitigation measures suiting the regional settings and also necessary to increase participation of the local community in the mitigation measures to make water conservation programs successful at the watershed level. It is also significant, as a part of a mitigation measure, to select suitable sites for artificial recharge structures for water conservation activity to get maximum output in the form of augmented groundwater (Kaliraj et al. 2015). This attempt requires site-specific knowledge regarding topography, geology, geomorphology, hydrology, climate, and other collateral factors. Geoinformatics has been used to process such spatial and non-spatial site-specific data and generate a map for artificial recharge potential zones and locating artificial recharge structures, suitable for a site with its type and location. It is also important to suggest methods for an increase in local community participation in activities for enhancement in groundwater recharge, which can effectively be used for efficient implementation of water conservation schemes at the local level.

For a mitigation plan, it is necessary to identify the locations for the artificial recharge structures along with appropriate artificial recharge structures suitable to site conditions considering the nature of various controlling factors like slope, geomorphology, and land use land cover. Combined use of remote sensing and GIS can be proved as a potent technique for planning of recharge structures. Artificial recharge is the most influential management tool to assure the sustainability of groundwater resources. There is a need to shift from the approach of broad-based aquifer systems or groundwater reservoirs to localized thematic information mapping. This can be targeted by the use of different methods wherein data analytics play a vital role. Number of models can be derived using tremendous data available on various themes, may be based on different assumptions or based on the conclusions drawn from the past studies. Groundwater observation, planning, and utility need to be changed from its old aged views to a newer version of ultramodern dimension, geospatial platforms, computing, and modeling, framed to tackle the issues of groundwater.

Artificial recharge systems are engineered systems where surface water is put on or in the ground for infiltration and subsequent movement to aquifers to augment groundwater resources (Patil and Lad, 2021). Artificial recharge has commonly referred to as the enhancing groundwater recharge over and above the natural recharge. Methods of artificial recharge are direct and indirect recharge, which consists of a water spreading method, induced recharge, injection method, percolation ponds, check dams, continuous contour trench, and nalla bunds. A wide range of techniques is in use to recharge the groundwater artificially. The selection of the suitable technique depends upon the hydrological framework of that particular region.

It is observed during many studies that the selection of a specific recharge method is mainly governed by local conditions and also by the techno-economic viability and social acceptability of such scheme. Suitability of a particular method is based on the nature and geometry of the aquifer system, its geo-hydrological parameters, quality of source water, and proposed use of the groundwater. Similarly, identification of an appropriate location for artificial recharge structure is essential from not only a construction aspect but also from operation and maintenance aspect. By superimposing a drainage map over an artificial groundwater recharge zones map, suitable locations for artificial recharge structures are identified and marked on the map using GIS tools. Finally, a map showing streams on groundwater recharge potential zones for the region can be generated, which is useful to locate the artificial recharge structures. Accordingly, spatial analysis is to be conducted in a GIS environment using weighted overlay analysis and a map for artificial groundwater recharge zones can be generated. Further, the study for choosing appropriate artificial recharge structures is to be conducted and appropriate recharge structures like percolation tanks and check dams can be selected, which can prove as most suitable for hydro-geomorphological settings of the region.

While implementing any of the water conservation measures, as a part of mitigation plan for enhancement in groundwater recharge by artificial recharge structures, selection of artificial recharge sites assumes great importance. Analyzing the geo-hydrological settings of locations from the available data as mentioned above, appropriate locations for artificial structures can be well identified, which are most suitable for enhancement in groundwater recharge. The structures commonly used are bench terracing, contour bunds, gully plugs, nalla bunds, check dams, and percolation ponds. The water stored in these structures is mostly confined to the stream course and the height is normally less than two meters. These structures are designed based on stream width and excess water is allowed to flow over the wall. In order to avoid scouring from excess runoff, water cushions are provided on the downstream side. To harness maximum non-committed surplus runoff in the stream, a series of such suitable recharge structures can be constructed at appropriate locations for artificial groundwater recharge on a regional scale (Patil and Lad, 2021).

6.5.4 WATERSHED MANAGEMENT

A watershed is an area of land that drains or sheds water into a specific waterbody. Water divides or the ridge line separates one watershed from another by a natural boundary. Watershed management is the process of creating and implementing plans, programs, and projects to sustain and enhance watershed functions that affect the plant, animal, and human communities within the watershed boundary. The strengths and weaknesses in the planning and implementation of water conservation works are to be identified, so that these strengths and weaknesses can be used further to prepare a management plan for watershed for not only enhancement in groundwater recharge but to have effective comprehensive watershed management for betterment of society through augmented water resources and thereby sustainable development. Most of the common components observed in the development project of any watershed are listed as follows:

- Soil conservation
- Massive tree plantation
- Water conservation
- Ancillary developments

Various activities under any of the watershed development projects mobilizes local villagers' participation to make such a project successful and sustainable, which includes:

- Establishment of good interaction with local community through entry program(s)
- Creation of awareness amongst local community regarding objectives of the project
- Planning for implementation at the local unit level, like a village, in dialogue with local villagers
- Formation of groups of stakeholders and committees for implementation of project
- Organizing capacity-building programs for the stakeholders' groups and project implementation committees
- Execution of various components through stakeholders' groups and project implementation committees
- Coordinating for active participation of local community in such watershed development projects

It is necessary to understand the groundwater regime before going to watershed management. Water occurring in geological formations is divided primarily into two zones. The upper one is the aeration zone and the lower one is the saturation zone. However, the unsaturated zone, lying between the aeration zone and saturation zone, particularly when the water table is not shallow but it is deep.

Plants require the proper type and quantity of soil for their growth and development. Therefore, the soil holds the significance in watershed management. The plants require nutrients to synthesize their food, along with light, air, heat, and water. Hence, the soil plays the role of a logical megastore for nutrients, which supports plants and acts as a prime factor in crop productivity. In the case of rapid loosening and loss of top soil due to erosion, it happens that the agricultural lands lose their crop productivity and rapidly deteriorates in quality due to improper or poor soil management. Therefore, it requires appropriate soil management system, which can reduce available soil water due to increase in run-off and density of soil.

Water harvesting is an easy method by which rainfall is collected, by minimizing the runoff and making an arrangement for reducing the rainwater flow by adopting various means for more infiltration, for future use. Artificial recharge structures store surplus non-committed run-off. Essential elements of any project like motive and objectives of the project, stakeholders, identification of problem, location of project, budgeting of the project, etc. are very important while planning a strategy for implementation of the project and terminology necessary to understand the project.

6.6 CASE STUDY ON IDENTIFICATION OF POTENTIAL GROUNDWATER RECHARGE ZONES AND SUITABLE LOCATIONS FOR APPROPRIATE ARTIFICIAL RECHARGE STRUCTURES USING REMOTE SENSING AND GIS TECHNOLOGY

A case study for delineating potential groundwater recharge zones and generation of zonation map with suitable locations of appropriate artificial groundwater recharge structures for water conservation work and mitigation plan for enhancement in groundwater recharge is discussed here briefly. The principal objective of study was to enhance groundwater recharge by developing a hydrological model that helps in identifying the spatio-temporal distribution of groundwater recharge in the upper Bhima basin in the Pune District. A model for enhancement of groundwater development and efficient groundwater management is developed using the knowledge obtained through the present study. The model also helps to suggest the sustainable groundwater recharge measures and also helps the decision maker to make timely decisions to enhance the groundwater recharge and save the precious natural resource for human beings.

From the outcomes of the study, the direct as well as indirect benefits, those that could be availed, can be estimated. Direct benefits from the program include the rise in the groundwater table, which helps to increase area for agriculture and also helps for availability of drinking water during the summer season. Indirect benefits include increase in the number of crops per year and also planting of a variety of crops and fruit bearing trees throughout the year, considering the assurance of availability of more water for agriculture. It impacts migration from village to city and, due to availability of job and business opportunities, the migration pace is slowed.

Spatial and non-spatial data were used to generate thematic maps using geoinformatics. These include primary data of pre-monsoon and post-monsoon observations of water levels for knowing water table fluctuation from key observation open dug wells and secondary data that include toposheets, satellite imagery, DEM, geological, geomorphological, soil, rainfall data, etc. Using the methodology as shown below (Figure 6.23) with geoinformatics techniques, various thematic maps are generated (Figure 6.25 (A – L)). Figure 6.24 shows a flowchart for a hydrological model.

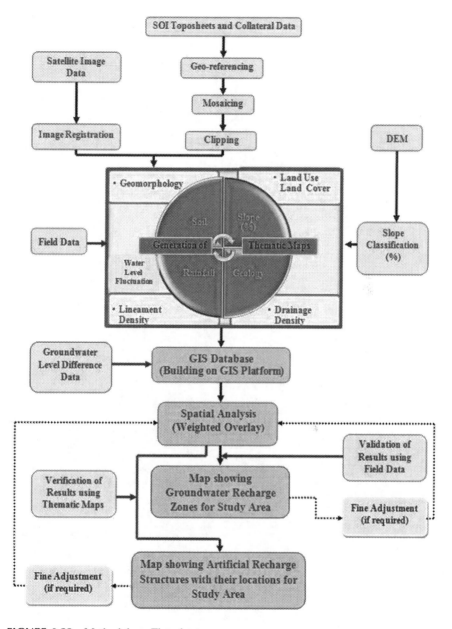

FIGURE 6.23 Methodology Flowchart.

These data have analyzed using ArcGIS tools to derive relevantly and to be used information. For spatial analysis, the influence factors of groundwater recharge are used to determine the spatio-temporal distribution of groundwater recharge and further to design the mitigation plan suggesting the mitigation measures for enhancement in groundwater recharge. The final output is obtained in the form of an

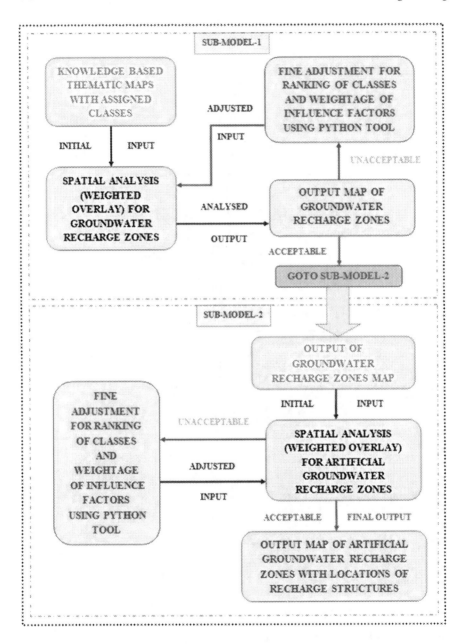

FIGURE 6.24 Flowchart for Hydrological Model.

artificial recharge potential zonation map along with the recharge structures suitably located in the study area (Figure 6.26).

During the study, various water conservation works carried out in the study area are investigated and following direct and indirect benefits are observed (Table 6.3).

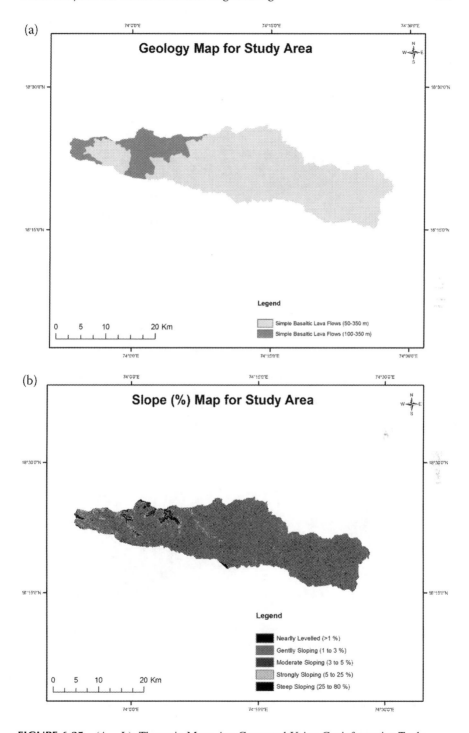

FIGURE 6.25   (A – L). Thematic Maps Are Generated Using Geoinformatics Tools.

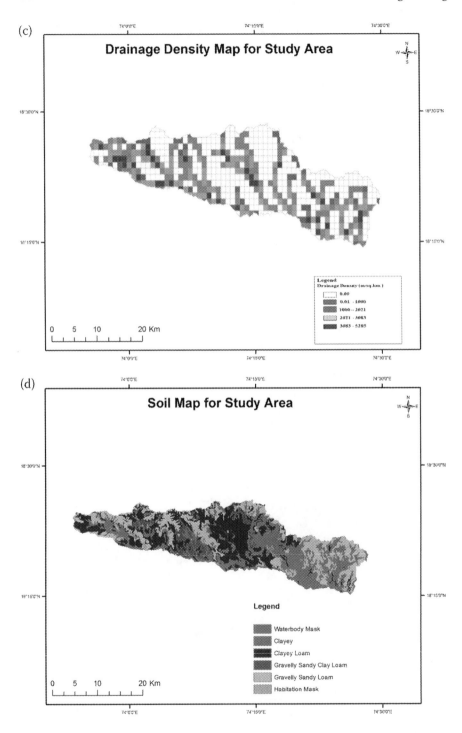

FIGURE 6.25   (*continued*)

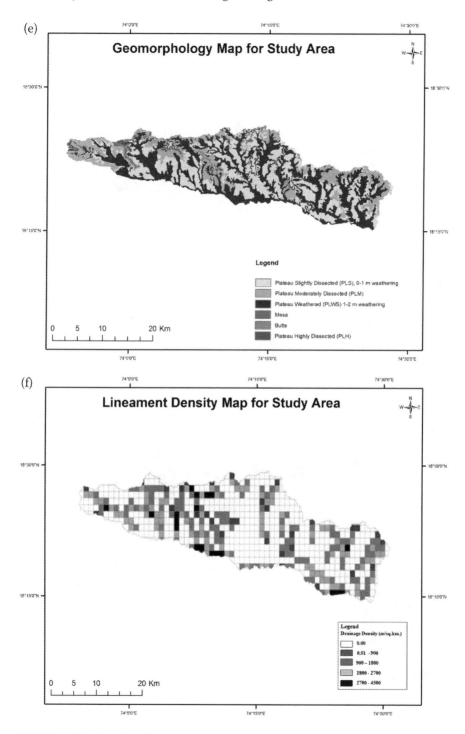

FIGURE 6.25 (*continued*)

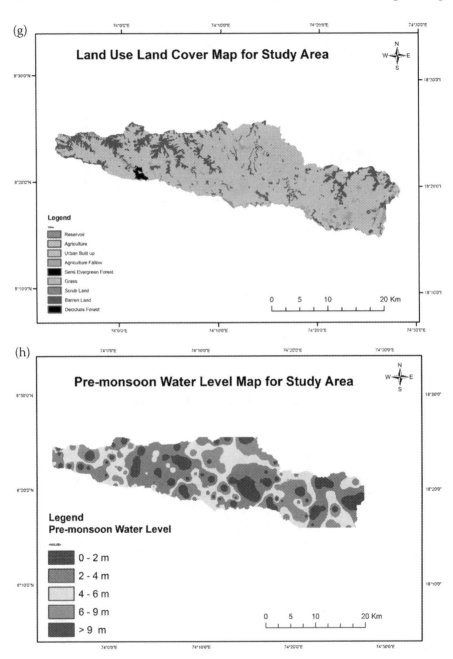

FIGURE 6.25   (*continued*)

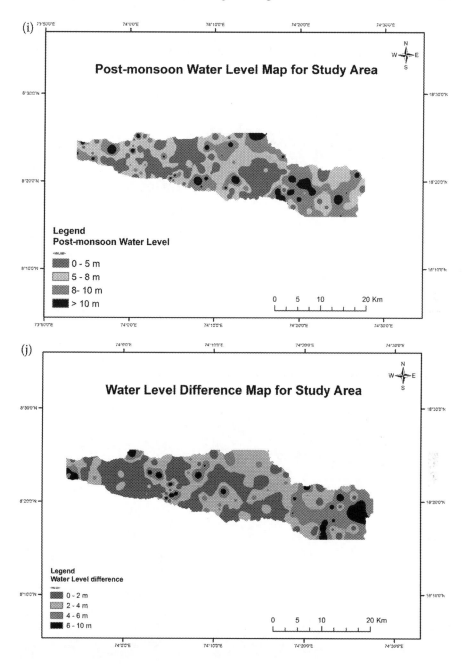

FIGURE 6.25 (*continued*)

(k)
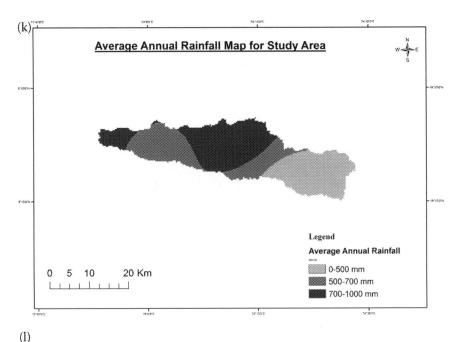

(l)
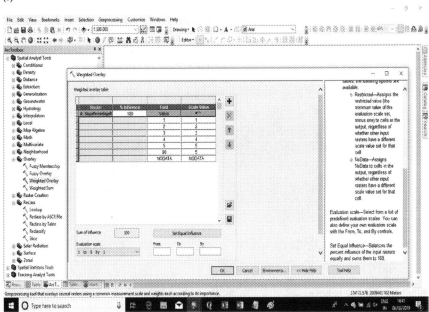

FIGURE 6.25   *(continued)*

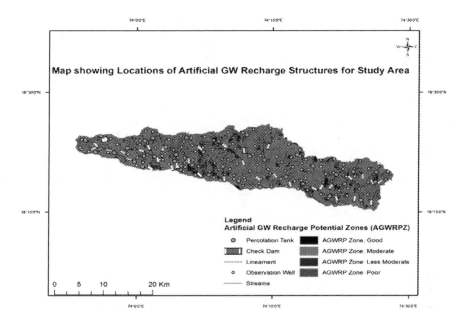

FIGURE 6.26 Map Showing Artificial Groundwater Recharge Structures.

TABLE 6.3
Water Conservation Works Are Investigated During Study

Sr. No.	Location	Works Carried Out with Public Participation	Works Completed (Physical)	Impacts of Works Completed Under J.S.A.
(1)	Sonwadi Supe	i. Nalla Deepening ii. CCTs iii. Cement Bunds iv. De-silting of Percolation Tanks v. Recharging of Dug Wells/Bore Wells	i. 5 km. ii. 25 Hect. iii. 4 Nos. iv. 2 Nos. v. 171 Nos./ 9 Nos.	i. Raised groundwater table ii. Increased vegetation cover iii. Additional crop per year iv. Increased availability of water for drinking purpose

In this case study data, is handled in a GIS environment and multicriteria-based weighted overlay analysis is carried out using a geoinformatics tool. As the spatial analysis takes more time for exploring more and more options with changing influence factors and their respective classes, it becomes cumbersome and to make it easy, a processing tool in Python is introduced, which saves time and also minimizes the efforts to make changes in the weightage and ranking during spatial analysis. It is, therefore, necessary to point out here that such a type of huge multi-faceted data may be easily handled using machine learning and artificial intelligence techniques for the future prediction of groundwater recharge zonation and further to find out the most proper location for appropriate recharge structure with ease and with multiple options to change inputs for obtaining the most accurate outputs.

REFERENCES

Adamowski, Jan, and Hiu Fung Chan. "A wavelet neural network conjunction model for groundwater level forecasting." *Journal of Hydrology* 407, no. 1-4 (2011): 28–40.

ADB. (2020). https://www.adb.org/sites/default/files/publication/614891/artificial-intelligence-smart-water-management-systems.pdf. [Retrieved: 20-09-2021].

Adnan, Rana Muhammad, Zhongmin Liang, Ahmed El-Shafie, Mohammad Zounemat-Kermani, and Ozgur Kisi. "Prediction of suspended sediment load using data-driven models." *Water* 11, no. 10 (2019): 2060.

Adusei, Yvonne Yeboah, Jonathan Quaye-Ballard, Albert Amatey Adjaottor, and Alex Appiah Mensah. "Spatial prediction and mapping of water quality of Owabi reservoir from satellite imageries and machine learning models." *The Egyptian Journal of Remote Sensing and Space Science* (2021).

Agrawal, D., Bernstein, P., Bertino, E., Davidson, S., Dayal, U., Franklin, M., and Widom, J. "Challenges and Opportunities with Big Data". A community white paper developed by leading researchers across the United States. Computing Research Association, Washington. (2012).

Ahmad, Sajjad, and Slobodan P. Simonovic. "System dynamics modeling of reservoir operations for flood management." *Journal of Computing in Civil Engineering* 14, no. 3 (2000): 190–198.

Ali, Anwaar, Junaid Qadir, Raihanur Rasool, Arjuna Sathiaseelan, Andrej Zwitter, and Jon Crowcroft. "Big data for development: applications and techniques." *Big Data Analytics* 1, no. 1 (2016): 1–24.

Amaranto, Alessandro, Francisco Munoz-Arriola, D.P. Solomatine, and Gerald Corzo. "A spatially enhanced data-driven multimodal to improve semi seasonal groundwater forecasts in the High Plains aquifer, USA." *Water Resources Research* 55, no. 7 (2019): 5941–5961.

Anamika Tirkey, Shalini, Mili Ghosh, and A.C. Pandey. "Soil erosion assessment for developing suitable sites for artificial recharge of groundwater in drought prone region of Jharkhand state using geospatial techniques." *Arabian Journal of Geosciences* 9, no. 5 (2016): 362.

Bejarano, Gissella, Mayank Jain, Arti Ramesh, Anand Seetharam, and Aditya Mishra. "Predictive analytics for smart water management in developing regions." In 2018 IEEE International Conference on Smart Computing (SMARTCOMP), pp. 464–469. IEEE, (2018).

Bernard, Marr. *Big Data: Using SMART big data, analytics and metrics to make better decisions and improve performance*. John Wiley & Sons, 2015.

Bhattacharya, B., R.K. Price, and D.P. Solomatine. "Data-driven modelling in the context of sediment transport." *Physics and Chemistry of the Earth, Parts A/B/C* 30, no. 4-5 (2005): 297–302.

Bhowmick, P., and Sivakumar, V. "Geoinformatics in Water Resource Management - GIS based Fuzzy Modelling approach to identify the suitable sites for Artificial Recharge of Groundwater, *LAP Lambert Academic Publishing* - ISBN: 978-3-659-58641-5, 2014.

Blaas, Meinte, Changming Dong, Patrick Marchesiello, James C. McWilliams, and Keith D. Stolzenbach. "Sediment-transport modeling on Southern Californian shelves: A ROMS case study." *Continental Shelf Research* 27, no. 6 (2007): 832–853.

Brabham, Daren C. "Crowdsourcing as a model for problem solving: An introduction and cases." *Convergence* 14, no. 1 (2008): 75–90.

CH, 2021, Data analytic course resources https://www.coursehero.com/. [Retrieved: 11-07-2021].

Chandarana, P., and M. Vijayalakshmi. "Big data analytics frameworks". In 2014 International Conference on Circuits, Systems, *Communication and Information Technology Applications (CSCITA)* (pp. 430–434). IEEE. 2014.

Chen, CL Philip, and Chun-Yang Zhang. "Data-intensive applications, challenges, techniques and technologies: A survey on Big Data." *Information Sciences* 275 (2014): 314–347.

Cui, Xuezhu, and Xuetong Wang. "Review on studies of urban spatial behavior and urban planning from the perspective of big data." In *ICCREM 2015*, pp. 521–531. 2015.

De, Arkajyoti, and Surya Prakash Singh. "Analysis of fuzzy applications in the agri-supply chain: A literature review." *Journal of Cleaner Production* 283 (2021): 124577.

Dogan, Emrah, Bülent Sengorur, and Rabia Koklu. "Modeling biological oxygen demand of the Melen River in Turkey using an artificial neural network technique." *Journal of Environmental Management* 90, no. 2 (2009): 1229–1235.

DSC, 2021, Water Management Analytics - Insights and Intelligence - Data Science Central, https://www.datasciencecentral.com/profiles/blogs/water-management-analytics-insights-and-intelligence. [Retrieved: 04-07-2021].

Fell, J., Pead, J. and Winter, K. Low-cost flow sensors: Making smart water monitoring technology affordable. *IEEE Consumer Electronics Magazine* 8, no. 1 (2018): pp. 72–77.

Gaffoor, Zaheed, Kevin Pietersen, Nebo Jovanovic, Antoine Bagula, and Thokozani Kanyerere. "Big data analytics and its role to support groundwater management in the southern African development community." *Water* 12, no. 10 (2020): 2796.

Gandomi, Amir, and Murtaza Haider. "Beyond the hype: Big data concepts, methods, and analytics." *International Journal of Information Management* 35, no. 2 (2015): 137–144.

Guo, Hua-Dong, Li Zhang, and Lan-Wei Zhu. "Earth observation big data for climate change research." *Advances in Climate Change Research* 6, no. 2 (2015): 108–117.

Howe, Jeff. "The rise of crowdsourcing." *Wired Magazine* 14, no. 6 (2006): 1–4.

Huang, Mutao, and Yong Tian. "Prediction of groundwater level for sustainable water management in an arid basin using data driven models." In International Conference on Sustainable Energy and Environmental Engineering (SEEE), Bangkok. 2015.

Hussein, Salah Ali, Mariana Mocanu, and Adina Florea. "Analysis of Data Mining Tools Used for Water Resources Management in Tigris River." *Advanced Management Science* 3, no. 2 (2014): 1–10.

Ilieva, Rositsa T., and Timon McPhearson. "Social-media data for urban sustainability." *Nature Sustainability* 1, no. 10 (2018): 553–565.

Janusz, Starzyk A. "Motivated learning for computational intelligence." In *Computational modeling and simulation of intellect: Current state and future perspectives*, pp. 265–292. IGI Global, 2011.

Jenny, Hubert, Eduardo Garcia Alonso, Yihong Wang, and Roberto Minguez. "Using Artificial Intelligence for Smart Water Management Systems." (2020).

Kaliraj, Seenipandi, Nainarpandian Chandrasekar, and N.S. Magesh. "Evaluation of multiple environmental factors for site-specific groundwater recharge structures in the Vaigai River upper basin, Tamil Nadu, India, using GIS-based weighted overlay analysis." *Environmental Earth Sciences* 74, no. 5 (2015): 4355–4380.

Kanda, Edwin Kimutai, Emmanuel Chessum Kipkorir, and Job Rotich Kosgei. "Dissolved oxygen modelling using artificial neural network: a case of River Nzoia, Lake Victoria basin, Kenya." pp 1–7 *Journal of Water Security* 2 (2016).

Khan M., and Ayyoob M. "Big data analytics evaluation. *Int. J. Eng. Res. Comput. Sci. Eng. (IJERCSE)*. 2018; 5(2):25–28.

Khudair, Basim Hussein, and N.O. Al-Musawi. "Water quality assessment and total dissolved solids prediction using artificial neural network in Al-Hawizeh Marsh South of Iraq." *Journal of Engineering* 24, no. 4 (2018): 147–156.

Klemen, Kenda, Matej Čerin, Mark Bogataj, Matej Senožetnik, Kristina Klemen, Petra Pergar, Chrysi Laspidou, and Dunja Mladenić. "Groundwater modeling with machine learning techniques: Ljubljana polje aquifer." In *Multidisciplinary Digital Publishing Institute Proceedings*, vol. 2, no. 11, p. 697. 2018.

Kolhe, Mohan Lal, Pawan Kumar Labhasetwar, and H.M. Suryawanshi, eds. *Smart technologies for energy, environment and sustainable development: Select proceedings of ICSTEESD 2018*. Springer, 2019.

Kumar, Vaibhav, and M.L. Garg. "Predictive analytics: a review of trends and techniques." *Int. J. Comput. Appl* 182, no. 1 (2018): 31–37.

LA, 2021, Educational resources. www.logianalytics.com. [Retrieved: 11-07-2021].

Lee, Jae-Gil, and Minseo Kang. "Geospatial Big Data: Challenges and Opportunities", *Big Data Research*, (2015).

Li, Jun, Zhi He, Javier Plaza, Shutao Li, Jinfen Chen, Henglin Wu, Yandong Wang, and Yu Liu. "Social media: New perspectives to improve remote sensing for emergency response." *Proceedings of the IEEE* 105, no. 10 (2017): 1900–1912.

Mayer-Schönberger, Viktor, and Kenneth Cukier. *Big data: A revolution that will transform how we live, work, and think*. Houghton Mifflin Harcourt, 2013.

Milly, P.C.D. "Climate change: stationarity is dead." *Whither Water Management Science* 319 (2008): 573–574.

Mohamed, Mohamed Mostafa, and Samy Ismail Elmahdy. "Fuzzy logic and multi-criteria methods for groundwater potentiality mapping at Al Fo'ah area, the United Arab Emirates (UAE): an integrated approach." *Geocarto International* 32, no. 10 (2017): 1120–1138.

Mohan, S., and N. Ramsundram. "Data-mining models for water resource applications", *ISH, Journal of Hydraulic Engineering*, (2013).

Moriarty, Julia M., Courtney K. Harris, and Mark G. Hadfield. "A hydrodynamic and sediment transport model for the Waipaoa Shelf, New Zealand: Sensitivity of fluxes to spatially-varying erodibility and model nesting." *Journal of Marine Science and Engineering* 2, no. 2 (2014): 336–369.

Naghibi, Seyed Amir, Hamid Reza Pourghasemi, and Barnali Dixon. "GIS-based groundwater potential mapping using boosted regression tree, classification and regression tree, and random forest machine learning models in Iran." *Environmental Monitoring and Assessment* 188, no. 1 (2016): 1–27.

NC-NIC. 2021. https://ncert.nic.in/textbook/pdf/legy305.pdf. [Retrieved: 12-10-2021].

Nemati, Samira, Mohammad Hasan Fazelifard, Özlem Terzi, and Mohammad Ali Ghorbani. "Estimation of dissolved oxygen using data-driven techniques in the Tai Po River, Hong Kong." *Environmental Earth Sciences* 74, no. 5 (2015): 4065–4073.

Nourani, Vahid, Mehdi Komasi, and Akira Mano. "A multivariate ANN-wavelet approach for rainfall–runoff modeling." *Water Resources Management* 23, no. 14 (2009): 2877–2894.

Ohlhorst, Frank J. *Big data analytics: turning big data into big money*. Vol. 65. John Wiley & Sons, 2012.

Pareta, Kuldeep, and Shirish Baviskar. "Geospatial Modeling for Planning of Water Conservation Structures." *American Journal of Geophysics, Geochemistry and Geosystems. American Journal of Geophysics, Geochemistry and Geosystems* 6, no. 4 (2020): 102–119.

Patil, S.G. "Spatio-Temporal distribution of groundwater recharge in Upper Bhima Basin using Geoinformatics". Ph. D. Thesis, SPPU, 1–182. 2021.

Patil, Shivaji Govind, and Ravindra Krishnarao Lad. "Evaluation of Spatio-temporal Dynamics of Groundwater Recharge and locating Artificial Recharge Structures for Watershed in Upper Bhima Basin, Pune, India." *Journal of the Indian Society of Remote Sensing* (2021): 1–22.

Prasad, Pankaj, Victor Joseph Loveson, Mahender Kotha, and Ramanand Yadav. "Application of machine learning techniques in groundwater potential mapping along the west coast of India." *GIScience & Remote Sensing* 57, no. 6 (2020): 735–752.

Pulido-Calvo, I., J. Roldán, R. López-Luque, and J.C. Gutiérrez-Estrada. "Demand forecasting for irrigation water distribution systems." *Journal of Irrigation and Drainage Engineering* 129, no. 6 (2003): 422–431.

Qihui, Wu, Guoru Ding, Yuhua Xu, Shuo Feng, Zhiyong Du, Jinlong Wang, Keping Long. "Cognitive Internet of Things: A New Paradigm Beyond Connection", *IEEE Internet of Things Journal*, 2014.

Rajasekhar, M., G. Sudarsana Raju, Y. Sreenivasulu, and R. Siddi Raju. "Delineation of groundwater potential zones in semi-arid region of Jilledubanderu river basin, Anantapur District, Andhra Pradesh, India using fuzzy logic, AHP and integrated fuzzy-AHP approaches." *HydroResearch* 2 (2019): 97–108.

Reinsel, D., J. Gantz, and J. Rydning. 2018. The digitization of the world from edge to core. IDC White Paper.

Ridouane, Chalh, Zohra Bakkoury, Driss Ouazar, and Moulay Driss Hasnaoui. "Big data open platform for water resources management." In *2015 International Conference on Cloud Technologies and Applications (CloudTech)*, pp. 1–8. IEEE, 2015.

Rozos, Evangelos. "Machine learning, urban water resources management and operating policy." *Resources* 8, no. 4 (2019): 173.

Ryan Cheah Wei Jie, Cha Yao Tan, Fang Yenn Teo, Boon Hoe Goh, Yau Seng Mah. "A Review of Managing Water Resources in Malaysia with Big Data Approaches", *Emerald*, 2021.

Sagan, Vasit, Kyle T. Peterson, Maitiniyazi Maimaitijiang, Paheding Sidike, John Sloan, Benjamin A. Greeling, Samar Maalouf, and Craig Adams. "Monitoring inland water quality using remote sensing: potential and limitations of spectral indices, bio-optical simulations, machine learning, and cloud computing." *Earth-Science Reviews* 205 (2020): 103187.

Sahoo, S., T.A. Russo, J. Elliott, and I. Foster. "Machine learning algorithms for modeling groundwater level changes in agricultural regions of the US." *Water Resources Research* 53, no. 5 (2017): 3878–3895.

Salari, Marjan, Esmaeel Salami Shahid, Seied Hosein Afzali, Majid Ehteshami, Gea Oliveri Conti, Zahra Derakhshan, and Solmaz Nikbakht Sheibani. "Quality assessment and artificial neural networks modeling for characterization of chemical and physical parameters of potable water." *Food and Chemical Toxicology* 118 (2018): 212–219.

Schreffler, C.L., Drought-trigger ground-water levels and analysis of historical trends in Chester County, Pennsylvania: U.S. Geological Survey Water-Resources Investigations Report, (1997) 97-4113, 6 p.

Sharma, Sanjib, Michael Gomez, Klaus Keller, Robert E. Nicholas, and Alfonso Mejia. "Regional Flood Risk Projections under Climate Change." *Journal of Hydrometeorology* 22, no. 9 (2021): 2259–2274.

Shen, Chaopeng. "A transdisciplinary review of deep learning research and its relevance for water resources scientists." *Water Resources Research* 54, no. 11 (2018): 8558–8593.

Simon, H.A.H.A., and P. Langley "Applications of machine learning and rule induction". *Communications of the ACM*, 38(11), 54–64. ACM. doi:10.1145/219717.219768. 1995.

Singh, K.P., A. Malik, D. Mohan, S. Sinha. Multivariate statistical techniques for the evaluation of spatial and temporal variations in water quality of Gomti River (India)—A
case study. *Water Res.* 38 (2004): 3980–3992.

Singh, Manoranjan Kumar, L. Rakesh, and Aniket Ranjan. "Evaluation of the risk of drug
addiction with the help of fuzzy sets." *Journal of Bioinformatics and Sequence
Analysis* 2, no. 4 (2010): 47–52.

Sivarajah, U., M.M. Kamal, Z. Irani, and V. Weerakkody. Critical analysis of Big Data
challenges and analytical methods. *J. Bus. Res.* 70 (2017): 263–286.

Solomatine, Dimitri P., and Avi Ostfeld. "Data-driven modelling: some past experiences and
new approaches." *Journal of hydroinformatics* 10, no. 1 (2008): 3–22.

Sun, Z., and Y. Huo. The spectrum of big data analytics. *J. Comput. Inf. Syst.* 2019.

Talend, 2021, Data analysis, *Resource centre*. www.talend.com. [Retrieved: 11-07-2021].

Ticlavilca, A.M., and M. McKee, Multivariate Bayesian regression approach to forecast
releases from a system of multiple reservoirs, *Water Resources Management*, 25
(2011): 523–543. 10.1007/s11269-010-9712-y.

TALEND, 2021, https://www.talend.com/. [Retrieved: 16-09-2021].

Verma, A., X. Wei, and A. Kusiak. "Predicting the total suspended solids in wastewater:
A data-mining approach". *Eng. Appl. Artif. Intell.* 2013, 26, 1366–1372.

Weeser, B., S. Jacobs, P. Kraft, M.C. Rufino, L. Breuer. "Rainfall-Runoff Modeling Using
Crowdsourced Water Level Data", *Water Resources Research*, 2019.

Xingdong, Deng, Penghua Liu, Xiaoping Liu, Ruoyu Wang, Yuanying Zhang, Jialv He, and
Yao Yao. "Geospatial big data: New paradigm of remote sensing applications." *IEEE
Journal of Selected Topics in Applied Earth Observations and Remote Sensing* 12, no.
10 (2019): 3841–3851.

Xue, Zuo, Ruoying He, J. Paul Liu, and John C. Warner. "Modeling transport and deposition
of the Mekong River sediment." *Continental Shelf Research* 37 (2012): 66–78.

Yoon, Jin-Ho, Kingtse Mo, and Eric F. Wood. "Dynamic-model-based seasonal prediction of
meteorological drought over the contiguous United States." *Journal of
Hydrometeorology* 13, no. 2 (2012): 463–482.

Zadeh, L.A. "Fuzzy sets information and control". *Information Sciences.* 8, no. 3, (1965):
338–353.

Zaman, Ghavidel Sarvin Zad, and Majid Montaseri. "Application of different data-driven
methods for the prediction of total dissolved solids in the Zarinehroud basin."
Stochastic Environmental Research and Risk Assessment 28, no. 8 (2014): 2101–2118.

Zhang, Changkuan, and Tang Hongwu. "Advances in Water Resources and Hydraulic
Engineering", Springer Nature, 2009.

Zheng, Wu Yi, Mahmoud El-Maghraby, and Sudipta Pathak. "Applications of deep learning
for smart water networks." *Procedia Engineering* 119 (2015): 479–485.

Zheng, Feifei, Ruoling Tao, Holger R. Maier, Linda See, Dragan Savic, Tuqiao Zhang,
Qiuwen Chen et al. "Crowdsourcing methods for data collection in geophysics: State of
the art, issues, and future directions." *Reviews of Geophysics* 56, no. 4 (2018):
698–740.

Zhengfa, Chen, and Guifeng Liu. "Application of Artificial Intelligence Technology in Water
Resources Planning of River Basin." In *2010 International Conference of Information
Science and Management Engineering*, vol. 1, pp. 322–325. IEEE, 2010.

7 Data Analysis in Geomatics

S S Shahapure

7.1 INTRODUCTION

In geomatics and surveying, the measurements like distances of the survey lines their directions are taken and these measurements are used for mapping the surface of the Earth, waterways, and establishing official boundaries in land. As the formation of the Earth is close to the ellipsoid, surveying is classified as a plane and geodetic. In a plane, surveying the Earth is considered to be a plane area as the area to be mapped is small. The two basic well-known principles of working from full to part and checking the work with another method is always adopted. The main survey stations are selected in the area to be surveyed. The lines connecting these main stations are called main survey lines. Other stations like tie stations may also be established for collecting topographical details (Figure 7.1).

In surveying, the first step followed is the selection of locations for main survey stations. Lines joining these survey stations are called survey lines. A set of such survey lines is called the traverse. There are two types of traverses 1) open traverse (Figure 7.2a): For these traverses, the starting point and end points are different. These are generally used when a survey is to be carried out for areas whose width is very small compared to length e.g., highways, railways, canals, etc. To check the accuracy of these open traverses, astronomical observations are carried out. 2) Closed traverse (Figure 7.2b): For these traverses, the starting point and end points coincide. These are generally used when a survey is to be carried out for an area that is not in a small strip form e.g. city boundary, village boundary, etc. To check the accuracy, geometrical conditions are applied.

Depending on the equipment used, various types of surveying such as chain, compass, plane table, theodolite, and tacheometric are performed to collect field data. Except chain surveying, the area is divided in any type of polygon (closed traverse). In chain surveying, the area is divided only in triangles as only the linear measurements are observed in chain surveying. While mapping a closed traverse, the end station and the starting station should coincide. Similarly in a closed traverse, the summation of the latitudes (projection of survey line on north-south direction) and the summation of the departures (projection of survey line on east-west direction) should be equal to zero. Due to the errors in the field measurements, the traverse fails to close on paper while plotting. The distance with which the end station of a closed traverse survey fails to meet with the starting one is known as closing error or error of closure (Figure 7.3). To eliminate closing error, Bowditch's Rule or transit rule or modified transit rule is applied (Kanetkar T.P. and Kulkarni S.V.).

DOI: 10.1201/9781003316671-7

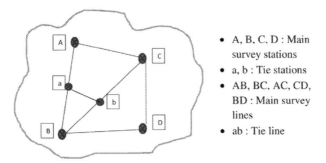

- A, B, C, D : Main survey stations
- a, b : Tie stations
- AB, BC, AC, CD, BD : Main survey lines
- ab : Tie line

FIGURE 7.1   Chain Survey (Kanetkar T.P. and Kulkarni S.V.).

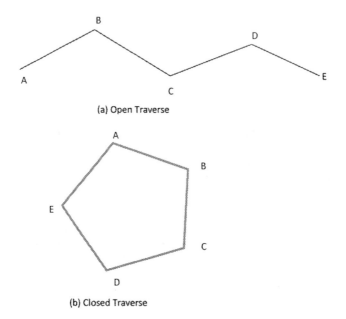

(a) Open Traverse

(b) Closed Traverse

FIGURE 7.2   Traverse (Kanetkar T.P. and Kulkarni S.V.).

In geodetic surveying, the effects of the curved shape of Earth and the refraction of light due to the atmosphere are accounted. The curved shape of earth and the refraction of light in the air affect the observations and require adjustment. These measurements need correction. Besides this, the precision required in geodetic surveying is more than plane surveying as a larger area of Earth's surface involved. The triangle formed by establishing stations on the surface of the Earth is not a plane triangle but a spherical triangle, as shown in Figure 7.4. For such a spherical triangle, the addition of all the angles of triangle is not exactly 180 degree but more than 180 degrees. This excess addition of angles of a triangle for a spherical triangle is called spherical excess. The magnitude of the excess is approximately one second for each 196 square kilometres area on the Earth ellipsoid (Chandra A.M.).

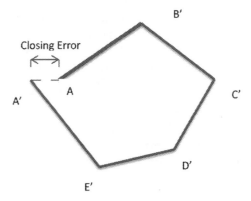

FIGURE 7.3   Closing Error (Kanetkar T.P. and Kulkarni S.V.)

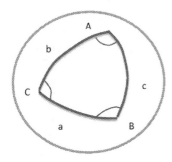

FIGURE 7.4   Spherical Triangle (Chandra A.M.).

In classical surveying, the linear, angular, and vertical measurements are observed with contemporary instruments. These observed measurements are to be corrected before using. With the advancement in using the electromagnetic radiations in measurements, the task of measurement has become simple. These measurements are required to be corrected and should satisfy geometric conditions as well. Nowadays, software, satellite imagery, sonar, 3D scanning, and drone technology are also used for measurements and their corrections. A lot of data through the field observations, satellite imagery, and positioning satellites are available. This data is required to be analyzed using different data analysis methods. The term *data analytics* means the process of examining data sets. It also helps in drawing conclusions about the information in data sets.

7.2   ADJUSTMENT OF SURVEY MEASUREMENT

The the field, observations made in surveying should be corrected before use.

7.2.1   INTRODUCTION

Survey measurements involve several operations such as centering, leveling, pointing, setting, and reading. Due to human limitations, imperfection in instruments and a certain amount of error are bound to occur in the measurements. The errors in the measured quantities should be minimized before they are used for computing other quantities. Errors are mainly classified as 1) blunders (mistake), 2) systematic error, and 3) random error. Mistakes or blunders can be detected by applying the geometric conditions e.g., summation of three angles of triangle is 180°0'0". Systematic errors can be eliminated by developing a formula e.g., $C_{cur} = 0.00785\ D^2$; C_{cur}: Correction for curvature in m; D: distance between instrument and staff in km. After removing the mistakes and systematic error, the errors that remain in the measurements are called residual error. These errors are minimized or adjusted and the value very close to true value is determined (Kanetkar T.P. and Kulkarni S.V.).

7.2.2 DEFINITIONS

1. True value: The value of a quantity which is free from all errors. It is not possible to eliminate all the errors. Therefore the true value of a quantity is indeterminate.
2. Observed value: It is the value obtained from the observation after eliminating mistakes and systematic errors. Observed value of a quantity may be an independent or conditional quantity.
3. Most probable value: It is the value which has more chances of being true than any other quantity.

True error = Observed value–True value
Residual error = Observed value–Most probable value

4. Normal equation: It is determined through following steps a) multiply each equation by the weight of observation and the coefficient of the unknowns b) add the equations thus determined.
5. Direct observation: The value of a quantity is measured directly.
6. Indirect observation: The value of a quantity is deduced from the measurements of other quantities.
7. Independent quantity: The value of an observed quantity is independent of other quantities.
8. Conditioned quantity: The value of an observed quantity is dependent upon the values of other quantities (Chandra A.M.).

7.2.3 ADJUSTMENT OF DIRECT AND INDIRECT OBSERVATIONS

All observations contain error. It is necessary to obtain most probable values of the observations (Chandra A.M.).

7.2.3.1 Weight of Observation (w)

A measurement of high precision will have small variance (σ^2). A measurement of low precision will have a large variance. The value of variance is inverse to that of the precision. However the weight of an observation is another measure of precision that is directly related to the precision of measurement. It is always represented in numbers. $w = k/(\sigma^2)$; where k is constant of proportionality (Chandra A.M.).

7.2.3.2 Laws of Weight

Following are the laws of weight based on method of least squares for error.

1. If n observations of unit (same) weight is used to find the arithmetic mean. Then the weight of arithmetic mean is n, number of observations.

Example: An < A was measured three times as below. If all the observations are of unit weight, what is the weight of arithmetic mean?

a. < A = 52°35'10"
b. < A = 52°35'12"
c. < A = 52°35'11"

Solution: The number of observations of unit weight n = 3.

$$Arithmetic\ mean\ =\ <A = 52°35'11"$$
$$The\ weight\ of\ Arithmetic\ mean\ =\ n = 3$$

2. If n observations of different weight used to find the weighted arithmetic mean. Then the weight of weighted arithmetic mean is equal to the sum of the individual weights.

Example: An angle A was observed three times as given below with their weights. What is the weight of the weighted arithmetic mean of the angle?

< A	Weight
65°21'7"	2
65°21'10"	3
65°21'8"	4

Solution:

$$weighted\ arithmetic\ mean = 65°21' + (7" \times 2 + 10" \times 3 + 8" \times 4)/9$$
$$= 65°21'8.\ 44$$
$$The\ weight\ weighted\ arithmetic\ mean = 2 + 3 + 4 = 9$$

3. The weight of the sum of the quantities added algebraically will be equal to the reciprocal of the sum of the reciprocal of the individual weight.

Example: Calculate the weight of (A + B) and (A − B) if the measured values and weights of A and B, respectively are

< A = 66°24′31″	weight (3)	< B = 61°22′42″	weight (4)

Solution:

< A + B = 127°47′13″	weight $\dfrac{1}{\frac{1}{3}+\frac{1}{4}} = \dfrac{12}{7}$
< A − B = 5°01′49″	weight $\dfrac{1}{\frac{1}{3}+\frac{1}{4}} = \dfrac{12}{7}$

4. The weight of the product of any quantity multiplied by a constant will be equal to the weight of that quantity divided by square of that constant.

Example: Calculate the weight of 3A, if < A = 20°32′24″   Weight (4)
Solution: 3A = 61°37′12″   Weight (4/3^2 = 4/9)

5. The weight of the quotient of any quantity divided by a constant will be equal to the weight of that quantity multiplied by square of that constant.

Example: Calculate the weight of A/4 if < A = 40°56′08″   Weight (2)
Solution: A/4 = 10°14′02″   Weight (2 ∗ 4^2 = 32)

6. The weight of a quantity remains the same if all the signs of the equation are changed or if a constant is added or subtracted.

Example: If weight of A + B = 29°34′20″ is 3, what is the weight of −A−B + 180.
Solution: −A−B+180 = 150°25′40″   weight = 3 (unchanged)
To determine the size and shape of a triangle, the minimum number of elements (n_o) required is three and one of them must be a side. Total number of elements (n) of triangle is six, i.e., length of three sides and three angles. If all the elements (n = 6) of the triangle are observed then there will be three redundant measurements (r = 3) r = n−n_o = 3. In surveying, redundant observations help in the detection of mistakes or blunders. The redundant observations need a method that can yield a unique solution. The least squares method is used for this purpose.

The least squares method of adjustment is based upon "The sum of the squares of the residuals (observational) must be minimum (Φ)".

$$\Phi = e_1^2 + e_2^2 + e_3^2 + \dots + e_n^2 = \sum_{i=1}^{n} e_i^2 = minimum\,(equal\ weights)$$

$$\Phi = w_1 e_1^2 + w_2 e_2^2 + w_3 e_3^2 + \dots + w_n e_n^2 = \sum_{i=1}^{n} w_i e_i^2 = minimum\,(unequal\ weights)$$

$e_1, e_2, e_3, \dots e_n$: Errors in observations (Chandra A. M.)

7.2.3.3   Direct Observation of Equal Weight

The most probable value of the directly observed of equal weights is the arithmetic mean of the observation (Chandra A.M.).

Example:

An angle A is measured six times with the following results:

$$A_1 = 50°20'\ 10''$$
$$A_2 = 50°20'\ 11''$$
$$A_3 = 50°20'\ 12''$$
$$A_4 = 50°20'\ 13''$$
$$A_5 = 50°20'\ 14''$$
$$A_6 = 50°20'\ 15''$$

Determine the most probable value of the angle by the method of least squares.

Solution:

1. Total number of observation (n) = 6.
2. The minimum number of observation to determine the angle (n_0) = 1.
   Let most probable value of the angle be A.
3. Redundant observation r = n − n_0 = 5.
   The number of conditions will be n' = 5 + 1 = 6.
4. The conditions are

$$A_1 + e_1 = A$$
$$A_2 + e_2 = A$$
$$A_3 + e_3 = A$$
$$A_4 + e_4 = A$$
$$A_5 + e_5 = A$$
$$A_6 + e_6 = A$$

Another condition from the least squares principle is

$$\emptyset = e_1^2 + e_2^2 + e_3^2 + e_4^2 + e_5^2 + e_6^2 = minimum$$
$$= (A - A_1)^2 + (A - A_2)^2 + (A - A_3)^2 + (A - A_4)^2 + (A - A_5)^2 + (A - A_6)^2$$
$$= minimum$$
$$\frac{d\emptyset}{dA} = 2(A - A_1) + 2(A - A_2) + 2(A - A_3) + 2(A - A_4) + 2(A - A_5) + 2(A - A_6) = 0$$
$$6A = A_1 + A_2 + A_3 + A_4 + A_5 + A_6$$
$$A = \frac{A1 + A2 + A3 + A4 + A5 + A6}{6} = 50°20'\ 13''$$

7.2.3.4 Direct Observations of Unequal Weights

The most probable value of directly observed quantity of observations of unequal weights is the weighted arithmetic mean of the observation (Chandra A.M.).

Example: An angle A is measured with an unequal weight and the following are the values.

$$
\begin{aligned}
< A_1 &= 61°24'11" \quad wt.\ 1 \\
< A_2 &= 61°24'12" \quad wt.\ 2 \\
< A_3 &= 61°24'13" \quad wt.\ 3 \\
< A_4 &= 61°24'14" \quad wt.\ 4 \\
< A_5 &= 61°24'15" \quad wt.\ 5 \\
< A_6 &= 61°24'16" \quad wt.\ 6
\end{aligned}
$$

Solution n = 6; n_0 = 1; r = 6 − 1 = 5.

$$\varnothing = w_1 e_1^2 + w_2 e_2^2 + w_3 e_3^2 + w_4 e_4^2 + w_5 e_5^2 + w_6 e_6^2 = minimum$$

$$= 1(A − A_1)^2 + 2(A − A_2)^2 + 3(A − A_3)^2 + 4(A − A_4)^2 + 5(A − A_5)^2 + 6(A − A_6)^2$$

$$= minimum$$

$$\frac{d\varnothing}{dA} = (A − A_1) + 2(A − A_2) + 3(A − A_3) + 4(A − A_4) + 5(A − A_5) + 6(A − A_6) = 0$$

$$A = 61°24'15.\ 04"$$

7.2.3.5 Indirect Observed Quantities Involving Unknown of Equal Weight

The most probable values of the quantities are expressed in terms of the observed quantities and the least squares criteria is applied to minimize the residuals (Chandra A.M.).

Example: Find the most probable values of A and B from the following observation with equal weights:

$$
\begin{aligned}
A' &= 53°29'12" \\
B' &= 20°25'14" \\
A' + B' &= 73°54'24"
\end{aligned}
$$

Solution: n = 3, n_0 = 2, r = 3 − 2 = 1.

$$
\begin{aligned}
A' + e_1 &= A = 53°29'12" + e_1 \\
B' + e_2 &= B = 20°25'14" + e_2 \\
(A' + B') + e_3 &= A + B = 73°54'24" + e_3
\end{aligned}
$$

From the least squares theory:

$$\emptyset = e_1^2 + e_2^2 + e_3^2 = minimum$$
$$= (A - 53°29'12'')^2 + (B - 20°25'14'')^2 + (A + B - 73°54'24'')^2$$
$$\frac{d\emptyset}{dA} = 2(A - 53°29'12'') + 2(A + B - 73°54'24'') = 0$$
$$\frac{d\emptyset}{dB} = 2(B - 20°25'14'') + 2(A + B - 73°54'24'') = 0$$
$$2A + B = 127°23'36'' = 127.3933°$$
$$A + 2B = 94°19'38'' = 94.3220$$

Solving:

$$B = 20.4203$$
$$= 20°25'13.32''$$
$$A = 53.4866° = 53°29'11.76''$$

7.2.3.6 Indirectly Observed Quantities Involving Unknowns of Unequal Weights

The most probable value of indirectly observed quantity with unequal weight can be determined in the following way (Chandra A.M.).

Example: Find the most probable value of the distance (a) from the following observations.

$$2a' = 300.600 \, m \quad wt - 1$$
$$3a' = 451.200 \, m \quad wt - 2$$
$$4a' = 600.500 \, m \quad wt - 3$$

Solution:
Let the residuals be e_1, e_2, and e_3.

$$\emptyset = w_1 e_1^2 + w_2 e_2^2 + w_3 e_3^2 = minimum$$
$$= 1(2a' - 300.6)^2 + 2(3a' - 451.200)^2 + 3(4a' - 600.5)^2 = minimum$$

Differentiating

$$\frac{d\emptyset}{da} = 70a - 10514.4 = 0$$
$$a = 150.206 \, m$$

7.2.4 ADJUSTMENT OF CLOSED TRAVERSE NETWORKS

Observation of closed transverses should satisfy geometric conditions. If the observation equation is accompanied by condition equations, the unknown residual errors of the observations may be deduced (Chandra A.M.)

Example: Three angles of a plane triangle are:

$$
\begin{aligned}
A &= 50°25'00'' \quad wt - 1 \\
B &= 60°35'00'' \quad wt - 2 \\
C &= 69°00'13'' \quad wt - 4
\end{aligned}
$$

Find the most probable values.

Solution:

$$
\begin{aligned}
A + B + C &= 180 \\
e_1 + e_2 + e_3 &= 180 - 50°25' - 60°35' - 69°00'13'' \\
e_3 &= -0°13'' - e_1 - e_2 \\
\emptyset &= w_1 e_1^2 + w_2 e_2^2 + w_3 e_3^2 \\
&= 2(e_1)^2 + 3(e_2)^2 + 4(-13'' - e_1 - e_2)^2 = minimum \\
\frac{d\emptyset}{de1} &= 2e_1 + 52'' + 4e_1 + 4e_2 = 0 \\
6e_1 + 4e_2 &= -52'' \\
\frac{d\emptyset}{de2} &= 3e_1 + 52'' + 4e_1 + 4e_2 = 0 \\
4e_1 + 7e_2 &= -52''
\end{aligned}
$$

Solving the above two simultaneous equations $e_1 = -6''$; $e_2 = -4''$; $e_3 = -3''$

<A = 50°24'54'';	<B = 60°34'56'';	<C = 69°00'10''

There is also another method, known as the method correction, to eliminate the condition equations.

Methods of correlates:

Correlates, the unknown multipliers (independent constants), are used for finding the most probable values of unknown parameters. The number of correlates is equal to the number of condition equations, excluding the one imposed by the least square principle (Chandra A.M.).

Example: Adjust the angle of a triangle ABC from the following data. Use a method of correlates:

$$
\begin{aligned}
<A &= 86°35'11.1'' \quad wt - 2 \\
<B &= 42°15'17.0'' \quad wt - 1 \\
<C &= 51°09'34.0'' \quad wt - 3
\end{aligned}
$$

Solution:

Step 1. It C_A, C_B, C_C correction in <A, <B, and <C. Total correction:

$$A + C_A + B + C_B + C + C_C = 180$$
$$C = C_A + C_B + C_C = -2.1'' \quad ...... \quad 1$$

Step 2. $\emptyset = w_A C_A{}^2 + w_B C_B{}^2 + w_c C_c{}^2 = minimum \quad .... \quad 2$

Step 3. Differentiate 1 and 2, i.e., condition equations.

$$\partial C = \partial C_A + \partial C_B + \partial C_C = 0 \quad .... \quad 3$$
$$\partial\emptyset = 2w_A c_A \partial C_A + 2w_B c_B \partial C_B + 2w_C c_C \partial C_C \quad .... 4$$

Step 4. Multiplying equation 3 by $(-\lambda)$ and adding the result to 4 we have

$$-\lambda(\partial C_A + \partial C_B + \partial C_C) + w_A C_A \partial C_A + w_B C_B \partial C_B + w_C C_C \partial C_C = 0$$
$$(w_A C_A - \lambda)\partial C_A + (w_A C_B - \lambda)\partial C_B + (w_C C_C - \lambda)\partial C_C = 0$$
$$w_A C_A - \lambda = 0; \quad w_B C_B - \lambda = 0; \quad w_C C_C - \lambda = 0$$

$$C_A = \frac{\lambda}{2}; \quad C_B = \frac{\lambda}{1}; \quad C_C = \frac{\lambda}{3}$$

$$\frac{\lambda}{2} + \frac{\lambda}{1} + \frac{\lambda}{3} = -2.1$$

$$\lambda = \frac{-2.1}{\frac{1}{2} + \frac{1}{1} + \frac{1}{3}} = -1.15$$

$$C_A = \frac{-1.15}{2} = -0.57$$

$$C_B = \frac{-1.15}{1} = -1.15$$

$$C_C = \frac{-1.15}{3} = -0.38; \quad sum = -2.10$$

<A = 86°35'11.1 − 0.57" = 86°35'10.53"

<B = 42°15'17.0 − 1.15" = 42°15'15.85"

<C = 51°0.9'34.0 − 0.38" = 51°09'33.62"

Sum = 180°00'00"

There are few alternative methods to determine the most probable values of the quantities such as forming the normal equations for the observed quantities, the method of difference which uses correction to simplify normal equations.

Geodetic surveying is carried by triangulation, trilateration triangulateration, or precise theodolite traversing. In triangulation, all the angles of the triangulation networks (Figure 7.5) are observed and a length of a base line is observed.

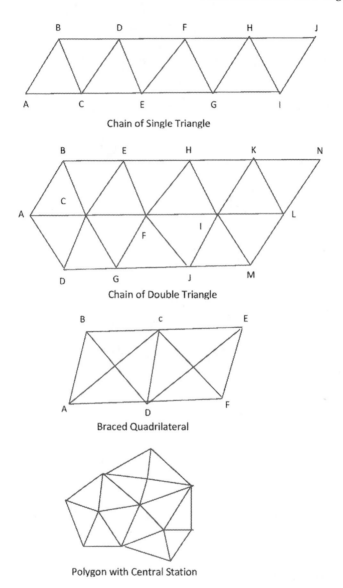

FIGURE 7.5 Triangulation Figures (Chandra A.M.).

Triangulation Adjustment

Observed angles are to be adjusted in triangulation networks (Figure 7.5). As it involves the adjustment of different types of figures with different geometric conditions, computation becomes extremely laborious in the absence of a digital computer with large memory.

The triangulation adjustment is divided into the following phases:

1. Angle adjustment
2. Station adjustment
3. Figure adjustment

It has been observed that even if the necessary geometrical conditions are satisfied, the outer sides may be discontinuous but parallel to their correct position. This problem is rectified by imposing an additional condition known as side condition. It is found from the fact that the product of the sines of the left-hand angles equals that of the sines of the right-hand angles. For a geodetic quadrilateral, by considering the logarithms of the sines, this condition becomes

$$\sum \log \sin \ (left \ hand \ angles) = \sum \log \sin \ (right \ hand \ angles)$$

7.2.4.1 Illustrative Example

The following eight angles of a geodetic quadrilateral PQRS were observed and adjusted for the closing errors of four stations P, Q, R, S. Adjust the angles by least squares method. The angles are corrected for spherical excess (Chandra A.M.) (Figure 7.6).

$\Theta_1 = 44°31'30''$	$\Theta_2 = 43°37'37''$	$\Theta_3 = 37°46'21''$	$\Theta_4 = 54°04'40''$
$\Theta_5 = 47°04'03''$	$\Theta_6 = 41°05'05''$	$\Theta_7 = 50°29'27''$	$\Theta_8 = 41°21'34''$

Solution: Let the correction to eight angles be $c_1, c_2, c_3 \ ..... \ c_8$ then

$$
\begin{aligned}
c_1 + c_2 + c_3 + c_4 + c_5 + c_6 + c_7 + c_8 &= 360° - \Sigma\Theta \\
&= 360° - 360°00'17'' \quad (7.1) \\
&= -17'' = D_1
\end{aligned}
$$

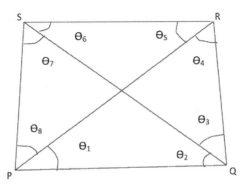

FIGURE 7.6 Geodetic Braced Quadrilateral (Chandra A.M.).

$$c_1 + c_2 - c_5 - c_6 = -(\Theta_1 + \Theta_2) + (\Theta_5 + \Theta_6)$$
$$= -(88°09'07'') + (88°09'08'') \qquad (7.2)$$
$$= +1'' = D_2$$

$$c_3 + c_4 - c_7 - c_8 = -(\Theta_3 + \Theta_4) + (\Theta_7 + \Theta_8)$$
$$= -(91°51'01'') + (91°51'01'') \qquad (7.3)$$
$$= 0'' = D_3$$

$$\Sigma \, log \, sin \, L = log \, sin \, \Theta_1 + log \, sin \, \Theta_3 + log \, sin \, \Theta_5 + log \, sin \, \Theta_7$$
$$= -0.1541455 - 0.2128742 - 0.1353960 - 0.1126512$$
$$= -0.6150669$$

$$\Sigma \, log \, sin \, R = log \, sin \, \Theta_2 + log \, sin \, \Theta_4 + log \, sin \, \Theta_6 + log \, sin \, \Theta_8$$
$$= -0.1611761 - 0.0916145 - 0.1823194 - 0.1799426$$
$$= -0.6150527$$

$$(\Sigma \, log \, sin \, L - \Sigma \, log \, sin \, R) \times 10^7 = (-0.61500669 + 0.6150527) \times 10^7$$
$$= -142$$
$$D_4 = -142 \, (to \, 7^{th} \, of \, decimal \, place)$$
$$Using \, f_1 = [log \, sin \, (\Theta_1 + 1'') - log \, sin \, (\Theta_1)]$$
$$\times 10^7 \, find \, f_1, \, f_2, \, ...f_8$$

To find the values of f_1, f_2, etc., using a pocket calculator, the following procedure may be adopted. Let $log \, sin \, (\Theta + 1'') = x$, and $log \, sin \, \Theta = y$.

$$Then \, taking \, x - y = 0.000000abc$$
$$f = 0.000000abc \times 10^7$$
$$i.\,e. = ab.c \, (to \, seventh \, place \, of \, decimal)$$

The example will make the determination of f more clear.

$$Log \, sin \, 44°31'31'' = x = -00.1541433$$
$$Log \, sin \, 44°31'30'' = x = -00.1541455$$
$$Therefore \, x - y = 0.0000022 \times 10^7$$

f = 22

$f_1 = 22; f_2 = 22; f_3 = 27; f_4 = 15; f_5 = 20; f_6 = 24; f_7 = 17; f_8 = 24$

The side condition gives $(\Sigma \log \sin L - \Sigma \log \sin R) \times 10^7 = D_4$.

Thus,

$$f_1 c_1 - f_2 c_2 + f_3 c_3 - f_4 c_4 + f_5 c_5 - f_6 c_6 + f_7 c_7 - f_8 c_8 = 21c_1 - 22c_2 + 27c_3 - 15c_4$$

$$+ 20c_5 - 24c_6 + 17c_7 - 24c_8$$

$$= -142 = D_4 \qquad (7.4)$$

From the least squares criterion, we have

$$\Phi = c_1^2 + c_2^2 + c_3^2 + c_4^2 + c_5^2 + c_6^2 + c_7^2 + c_8^2 = a\ minimum \qquad (7.5)$$

The procedure to determine the corrections in terms of correlates and the equations containing the correlates as unknowns is adopted. It includes steps, as explained in the method of correlates i.e., differentiating, multiplying with correlates ($\lambda_1, \lambda_2, \lambda_4$), and adding. The results are directly reproduced here.

$$c_1 = \lambda_1 + \lambda_2 + f_1 \lambda_4 \qquad (7.6)$$

$$c_2 = \lambda_1 + \lambda_2 - f_2 \lambda_4 \qquad (7.7)$$

$$c_3 = \lambda_1 + \lambda_3 + f_3 \lambda_4 \qquad (7.8)$$

$$c_4 = \lambda_1 + \lambda_3 - f_4 \lambda_4 \qquad (7.9)$$

$$c_5 = \lambda_1 + \lambda_2 + f_5 \lambda_4 \qquad (7.10)$$

$$c_6 = \lambda_1 - \lambda_2 - f_6 \lambda_4 \qquad (7.11)$$

$$c_7 = \lambda_1 - \lambda_2 + f_7 \lambda_4 \qquad (7.12)$$

$$c_8 = \lambda_1 - \lambda_3 - f_8 \lambda_4 \qquad (7.13)$$

Substituting $c_1, c_2, .. c_8$ in equations (7.1) to (7.5) and substituting $f_1, f_2, .. f_8$, we get the following:

$$8\lambda_1 + 0\lambda_4 = -17$$
$$4\lambda_2 + 3\lambda_4 = 1$$
$$4\lambda_3 + 19\lambda_4 = 0 \qquad (7.14)$$
$$0\lambda_1 + 3\lambda_2 + 19\lambda_3 + 3720\lambda_4 = -142$$

From the above equations, we get

$$\lambda_1 = -17/8 = -2.125$$
$$\lambda_2 = (1 - 3\,\lambda_4)/4 = 0.25 - 0.75\,\lambda_4$$
$$\lambda_3 = (-19/4)\lambda_4 = -4.75\,\lambda_4$$

Substituting the values of λ_1, λ_2, and λ_3 in equation 7.14, rearranging and clearing, we get

$$\lambda_1 = -2.125; \quad \lambda_2 = +0.279: \quad \lambda_3 = +0.185; \quad \lambda_4 = -0.039$$

Thus,

$$c_1 = -2.125 + 0.279 + 2 \times (-0.039) = -2.665''$$
$$c_2 = -2.125 + 0.279 - 22 \times (-0.039) = -0.988''$$
$$c_3 = -2.125 + 0.185 + 27 \times (-0.039) = -2.665''$$
$$c_4 = -2.125 + 0.185 - 15 \times (-0.039) = -1.355''$$
$$c_5 = -2.125 - 0.279 + 20 \times (-0.039) = -3.184''$$
$$c_6 = -2.125 - 0.279 - 24 \times (-0.039) = -1.468''$$
$$c_7 = -2.125 - 0.185 + 17 \times (-0.039) = -2.973''$$
$$c_8 = -2.125 - 0.185 - 15 \times (-0.039) = -1.374''$$
$$c_1 + c_2 + c_3 + c_4 + c_5 + c_6 + c_7 + c_8 = -17''$$
$$c_1 + c_2 - c_5 - c_6 = +1''$$
$$c_3 + c_4 - c_7 - c_8 = 0''$$

Thus, the most probable value of the angle of the angles are:

$$\Theta_1 + c_1 = 44°31'30'' - 2.665'' = 44°31'27.33''$$
$$\Theta_2 + c_2 = 43°37'37'' - 0.988'' = 43°37'36.01''$$
$$\Theta_3 + c_3 = 37°46'21'' - 2.993'' = 37°46'18.01''$$
$$\Theta_4 + c_4 = 54°04'40'' - 1.355'' = 54°04'38.65''$$
$$\Theta_5 + c_5 = 47°04'03'' - 3.184'' = 47°04'59.82''$$
$$\Theta_6 + c_6 = 41°05'05'' - 1.468'' = 41°05'03.53''$$
$$\Theta_7 + c_7 = 50°29'27'' - 2.973'' = 50°29'24.03''$$
$$\Theta_8 + c_8 = 41°21'34'' - 1.374'' = 41°21'32.63''$$
$$\Theta_1 + \Theta_2 + \Theta_3 + \Theta_4 + \Theta_5 + \Theta_6 + \Theta_7 + \Theta_8 = 360°00'00''$$
$$\Theta_1 + \Theta_2 - (\Theta_5 + \Theta_6) = 0$$
$$\Theta_3 + \Theta_4 - (\Theta_7 + \Theta_8) = 0$$

Example: Adjust the angles of the above geodetic quadrilateral by an approximate method.

Solution: The computation has been shown in Table 7.1. The values shown in the various columns have been computed as below.

Column 1 : Write $\Theta_1, \Theta_2, \Theta_3 \dots , \Theta_8$.
Column 2 : Enter the values of the observed angles against $\Theta_1, \Theta_2, \Theta_3 \dots , \Theta_8$.
Column 3 : The total correction for sum of angle is

$$(360° - \Sigma\Theta) = 360° - 360°00'17" = -17"$$
$$Correction\ to\ each\ angle = -17/8 = -2.125"$$

Enter the corrections for each angle.
Column 4: Enter the corrected angles after applying the corrections. Check that $\Sigma\Theta = 360°$.
Column 5: Determine the corrections for the opposite angle condition, as below:

$$\Theta_1 + \Theta_2 - (\Theta_5 + \Theta_6) = 88°09'2.75" - 88°09'3.75" = 1" (ignoring\ the\ sign)$$
$$Correction\ to\ each\ angle = \frac{1}{4} = 0.25"$$
$$\Theta_3 + \Theta_4 - (\Theta_7 + \Theta_8) = 0.00"\ the\ correction\ to\ these\ angles\ will\ be\ zero.$$

Column 6: Enter the corrected angles for the opposite angles conditions.
Column 7 and 8: Enter the values of (log sin Θ) for the values of Θ in column 6, ignoring the sign.
Column 9: Enter the values of f, which is difference for 1"of the angle for (log sin Θ) at the seventh place of decimal, i.e., the difference value $\times 10^7$.
Column 10: Enter the value of f^2.
Column 11: Enter the corrections for side equation. The corrections are computed as follows:

$$Correction = f/(\Sigma f^2) \times delta$$
$$delta = (\Sigma\ log\ sin\ L - \Sigma\ log\ sin\ R) \times 10^7$$
$$= (0.6150850 - 0.6150710) \times 10^7 (ignoring\ the\ sign.)$$
$$= 140.$$

The correction to the left angles will be negative to the right angles, since Σ log sin L is greater than Σ log sin R.

Apply the corrections given in column 11 to the angles given in column 6 and enter them in column 12. These are the final adjusted values of the angles. Check that thesum of the final angles is 360°

7.3   DATA ANALYSIS IN SATELLITE-BASED POSITIONING SYSTEM

A satellite system, orbiting at an altitude of 20,200 km, is being used for finding out the location (coordinates) on the surface of Earth.

TABLE 7.1
Adjustment of Quadrilateral by Approximate Method (Chandra A.M.)

θ	Observed Angle	Correction for Sum of all Angle	Corrected Angle	Correction for Sum of opp. Angles	Corrected Angles	Log Sin Left Angle	Log Sin Right Angle	f	f²	Side Equation Correction	Adjusted Angle
1	2	3	4	5	6	7	8	9	10	11	12
θ_1	44°31'30"	−2.125"	44°31'27.875"	+0.25"	44°31'28.125"	0.1541495		22	484	−0.828"	44°31'27.297"
θ_2	43°37'37"	−2.125"	43°37'34.875"	+0.25"	43°37'35.125"		0.1611802	22	484	+0.828"	43°37'35.953"
θ_3	37°46'21"	−2.125"	37°46'18.875"	00"	37°46'18.875"	0.2128800		27	729	−1.016"	37°46'17.889"
θ_4	54°04'40"	−2.125"	54°04'37.875"	00"	54°04'37.875"		0.0916179	15	225	+0.565"	54°04'38.440"
θ_5	47°04'03"	−2.125"	47°04'00.875"	−0.25"	47°04'00.625"	0.13454006		20	400	−0.753"	47°03'59.872"
θ_6	41°05'05"	−2.125"	41°05'02.875"	−0.25"	41°05'02.625"		0.1823252	24	576	+0.903"	41°05'03.528"
θ_7	50°29'27"	−2.125"	50°29'24.875"	00"	50°29'24.875"	0.1126549		17	289	−0.640"	50°29'24.235"
θ_8	41°21'34"	−2.125"	41°21'31.875"	00"	41°21'31.875"		0.1799477	24	576	+0.903"	41°21'32.778"
Σ	360°00'17"		360°00'00"		360°00'00.00"	0.6150850	0.6150710		3720		360°00'00.00"

7.3.1 INTRODUCTION

Initially, the U.S. Department of Defense developed the navigation satellite timing and ranging global positioning system (NAVSTAR GPS) and by the mid-nineties is was an all-weather high-accuracy radio navigation and positioning system. It has revolutioned the field of modern surveying navigation and mapping. In geodetic fields, GPS is likely to replace most of the other techniques. The GPS includes a satellite in near-circular orbits. These are at about 20,200 km altitude. This system gives full coverage. A inimum of four satellites are available to the user at any place on Earth. The signals transmitted by a minimum of four satellites simultaneously are used to determine the geometric position (latitude, longitude, and height) co-ordinated universal time. It provides a velocity vector with higher accuracy. This system has an advantage in economy and time over any other techniques available. This technique solves the numerous shortcomings of terrestrial surveying methods like the requirement of intervisibility of survey stations, dependability on weather, difficulties in night observation, etc. These advantages make GPS the encouraging surveying technique of the future. The raw GPS data includes many errors like satellite orbit error, multipath error, clock error, atmospheric error, etc.

From GPS observations, give the Cartesian rectangular coordinates x,y,z in an ECEF (Earth centered Earth fixed) global reference system. The transformation of coordinates from the global system to the local system is required. The GPS co-ordinates are in the global world geodetic system 1984 (WGS 84). This system is developed by the Defense Mapping Agency (DMA) of the U.S.A. These are to be transformed to the local data of particular country, e.g., Everest Ellipsoid in India. These transformations of coordinates involve several transformation parameters. These transformation parameters must be estimated. The value of these parameters evaluated by DMA need to be refined by rigorous computations for accuracy in coordinates.

The conversation from geodetic coordinates (latitudes, longitudes, and height) to the grid coordinates on Indian topographical gridded maps (easting, northing, and height) is required. It includes the transformation from the local geodetic system to the grid system superimposed on the map projection. Standard computer programs are available for this transformation.

Ellipsoidal height (height of the observation point above the reference (He) el-lipsoid) is deduced from GPS observation. The geodetic height (height above the geoid, commonly termed mean sea level (H) (MSL)) of a point is the geodal height that is calculated as follows. These two are related by a simple equation.

MSL Height (H) = Ellipsoidal height (He) = Geoidal undulations (N). The geodal undulation is derived from astro-geodetic or gravimetric data (Fell, P.J. (1980); Hoffman-Wellenhof, B et. al. (1997); Kulkarni, M.N. (2000)).

7.3.2 FEATURES OF GPS DATA PROCESSING

Following are some of the common features of GPS data processing for high-precision geodesy (Geoffrey Blewitt, 1998).

User Input Processing
- Operating system interface
- Interactive user control
- Automation (batch control defaults, contingency rules, etc.)
- Help (user manual, online syntax help, online module guide, etc.)

Data Preprocessing
- GPS observation files site database, Earth rotation data, satellite ephemerides, surface meteorological data, water vapor radiometer data
- Formatting
- Tools (satellite removal, data windowing, concatenation, etc.)
- Editing (detecting and removing outline and cycle slips)
- Thinning (data decimation, data smoothing)
- Data transformation (double differencing, ionosphere- free combination, etc.)
- Ambiguity initialization (and possible resolution)

Observation Models
- Normal values for model parameters
- (Observed- computed) observations and partial derivatives
- Orbit dynamics and satellite orientation
- Earth rotation and surface kinematics
- Media propagation (troposphere, ionosphere)
- Clocks
- Relativistic corrections (clocks, spacetime curvature)
- Antenna kinematics
- Phase modeling (phase center variation, polarization, cycle ambiguity)

Parameter Estimation
- Parameter selection
- Stochastic model and a priori constraints
- Inversion (specific algorithms, filtering, blocking techniques, etc.)
- Residual analysis (outline, cycle slips) and re-processing
- Sensitivity analysis (to unestimated parameters)

Solution Processing
- A priori constraints
- Ambiguity resolution
- Solution combination and kinematic modeling
- Frame projection and transformation tools
- Statistics (formal errors, repeatability, in various coordinate systems, etc.)

Output Processing
- Archive solution files
- Information for the user
- Export formatting (RINEX,SINEX,IONEX,etc.)
- Presentation formatting (e.g., graphics)

The equivalence of pseudorange and carrier phase can be used to develop data processing algorithms. The equivalence of stochastic and functional model leads to different methods of estimating parameters (Geoffrey Blewitt, 1998).

7.3.3   GPS Data Processing

GPS data consists of either pseudo-range or phase differences or both. GPS data gives the three-dimensional position of the receiver antenna at the instance of observation (either in cartesian co-ordinates x,y,z or latitude, longitude, and height) and the precise time. Other parameters such as atmospheric (tropospheric and ionospheric) delays, clock errors, satellite orbital elements, earth rotation parameters, and model parameters, etc. are also estimated. Based on accuracy desired, hardware available, surveying technique used to collect the data, etc., which depends on the choice of the software and processing technique to be used. The GPS processing software is classified into two categories. The receiver-specific vendor-supplied software is available. The GPS data collected using different makes of the receivers is in different formats. The facility of converting all these to a common internationally accepted ASCII format called RINEX (Receiver Independent Data Exchange Format) is available. It is possible to process any kind of data using the scientific software, for achieving higher accuracy (Kulkarni, 1999). The GPS observations are post-processed in relative mode using suitable software and accurate results are obtained. For most surveying applications, use of the vendor-supplied GPS data processing software is sufficient to yield the required accuracy. However, for high-precision scientific applications, special scientific data processing software being used worldwide includes the BERNESE developed by the University of Bern (Hugentobler et al., 2001); GAMIT developed by MIT, USA; GEODYN by NASA; GIPSY by JPL; TOPAS, Germany; GPSOBS by ESA; TEXGAP by the University of Texas; GEPHARD by GFZ, Germany; etc. Of these, BERNESE software is used by research institutes/organizations in India, including Survey of India, IIG, IIT Bombay CMMACS WIHG, NGRI, etc. and GAMIT is also being used in more recent times.

GPS has applications ranging from the mm-level high-precision Geodesy to several meter level navigation positioning. It has applications in the establishment of high-precision zero-order Geodetic National Survey Control Network of GPS stations, determining precise Geoid using GPS data, crustal movement of the order of few cm/year can be monitored using GPS method, and many more.

7.3.3.1   Applications

In February 2001, the GPS team from IIT Bombay carried out GPS observations to understand the post-earthquake crustal deformation pattern in the Bhuj region of Gujrat immediately after the 26th January 2001 earthquake. The object was also to monitor any future crystal dynamics in this region. GPS field obervations at 17 stations were repeated in February 2002. There existed several geodetic stations of the Great Trigonometrical (GT) Triangultion Network of India, established during the mid-nineteenth century in the Bhuj region. GPS observations at these stations would yield valuable data about crustal deformations in the region due to various causes, including the earthquakes of 1919 and 2001. A total of 17 stations have been observed using 4000 SSI Trimble dual frequency geodetic GPS receivers. The observations were carried out in four campaigns with 48 hours of continuous observations at every station. In February 2002, the team carried out repeat observations of all the stations in the Bhuj region. The baseline lengths and the respective root mean square (RMS) values of baselines between the stations were estimated using GPS survey TGO and Bernese software.

7.4   GEOSPATIAL ANALYSIS

Geospatial analysis includes the collection display and manipulation of imagery, GPS, and historical data.

7.4.1   REMOTE SENSING

Remote sensing is a process of obtaining information about an object, area, or phenomena through the analysis of data by a device without being in contact with the object, area, or phenomena being studied. Remote sensing is classified in two categories.

1. Active remote sensing
2. Passive remote sensing (use of solar radiations)

Most of the remote-sensing methods make use of the reflected infrared band, thermal infrared band, and microwave portion of the electromagnetic spectrum. Sensors used in remote sensing are mechanical devices that collect information usually in storable form about objects or scenes while at some distances from them. The data products used in remote sensing are photographs, mosaics, orthophoto, and satellite imagery. Remotely sensed data can be analyzed and interpreted either by visual method or digital method. Analyzing and interpreting the digital data is called digital image processing (Chandra A.M.).

7.4.2   GEOGRAPHIC INFORMATION SYSTEM (GIS)

A geographic information system is a computer-based system that manipulates the spatial and non-spatial data in digital form. GIS is a powerful set of tools for collecting, storing, retrieving, updating, manipulating, analyzing, and displaying all forms of geographically referenced information in some form for set of purposes. The major components of GIS are computer hardware, software modules, and a proper organizational context (Chandra A.M.).

7.4.3   DIGITAL IMAGE PROCESSING

Digital image processing is a vital part of remote-sensing operation. All digital imagery must be processed before use in the majority of applications. The image has to be displayed on a screen or as a photographic hard copy for visual interpretation. In many cases, more complex processing is needed in order to produce a final product for use.

The digital image processing is processing and analyzing the data stored in CCT- or PC-based/mainframe computer system using image processing algorithms. The technique has the following broad operations:

1. Image rectification and restoration: It is required to correct the distorted or degraded image data. This processing is required to correct for geometric distortions, to calibrate the data radiometrically and to eliminate noise present in the data.
2. Image enhancement: It is a technique for increasing visual distinction of features.
3. Image classification: Visual analysis is replaced with quantitative techniques for automating the identification the identification of objects in an image. Normally, multispectral image data that is involved in statistically based decision rules are used for determining the land cover classification for spectral radiance observed in the data. The classification is also based on geometrical shapes, size, and pattern present in the image called spatial pattern recognition. The object is to categorize all pixels in a digital image into one of several land cover classes.

Data merging: Data merging technique are applied to combine data (Chandra A.M).

7.4.3.1 Application of Geospatial Analysis

Rao and Rao (1988, 1997) used parameters obtained from remote-sensing (Landsat) data for simulation of a runoff from a watershed. Lhomme et al. (2004) used a GIS-based geomorphological routing model to simulate to a physically based routing model. Eldho (2006) added features of a finite element method to a geographical information system to take into account the spatial variation of data of the watershed. Roux and Dartus (2006) used remotely sensed data for getting a flood hydrograph. Dutta (2007) prepared a geographic information system–based comprehension raster database of different spatial data layers and used in the hydrodynamic modeling for flood inundation phenomena. Reddy et al. (2007) developed a kinematic wave–based distributed watershed model using a finite element method, geographical information system, and remote sensing for the runoff simulation of a small watershed.

7.4.3.2 Case Study

Shahapure S.S et al. (2011) have applied a finite element model to an urban coastal catchments located at Kalamboli, Navi Mumbai, India. The said catchment is divided into 40 strips based on flow direction and drainage pattern using the editor toolbar of the ArcGIS. These strips are divided into 181 one-dimension FEM grids, as shown in Figure 7.7.

The 35th strip drains into the channel through the detention pond of area 1.5 ha. The contour map of the catchment was georeferenced and the contour lines were digitized in the editor toolbox of the ArcGIS software and a polyline shape file was created. The data of spot level were obtained and used to create a point shape file. A

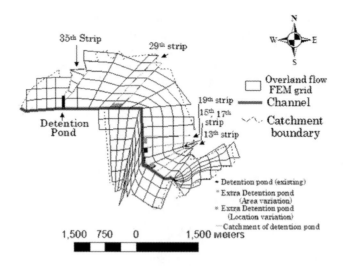

FIGURE 7.7 FEM Grid (Shahapure S.S et al., 2011).

catchment boundary was delineated. The digital elevation model (DEM), a digital representation of ground surface topography, was prepared using ArcGIS. A satellite image of the catchment area was captured with IRS-P6 LISS III with a spatial resolution of 23.5 m was used. A land use/land cover map was derived from this image using the ERDAS IMAGINE software, as shown in Figure 7.8.

Manning's roughness factor for each other overland flow element was determined using the reclassify and zonal statistics toolbar in ArcGIS.

Shahapure S S (2010) determined the slope for overland flow using GIS from DEM of a catchment located at Nerul, Navi Mumbai, India. Manning's roughness factor was also determined from land use classification using remote-sensing data.

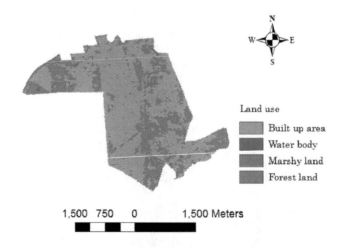

FIGURE 7.8 Land Use Land Cover (Shahapure S.S et al., 2011).

7.5 CONCLUSION

In geomatics and surveying, a lot of data is available through field observation, satellite imagery, and positioning satellite. This data is required to be corrected and analyzed using different data analysis methods. The field observation may be taken directly or indirectly with different precision and involves errors. It is necessary to find the most probable values for the observation that will satisfy the geometric conditions using mathematical techniques, e.g., least squares method. Data from the positioning satellites is available through the GPS receiver. There are many parameters involved in GPS data. Many types of GPS data analysis software are available with which very high precision in positioning data can be achieved. Due to the high accuracy and all-weather operation offered by GPS, it has numerous applications in many fields, ranging from the mm-level high precision Geodesy to the several-meter level navigational positioning such as establishment of high-precision zero-order Geodetic National Survey Control Network of GPS stations, determining precise Geoid using GPS data, crustal movement of the order of few cm/year can be monitored using the GPS method, and many more. The raw data available through remote sensing is in the form of photography, mosaics, ortho-photo, and satellite imagery. This raw data stored is processed and analyzed using image processing algorithms. The processed remote-sensing data can be used by hydrologists, agriculturists, geologists, town planners, and environmentalists.

REFERENCES

Blewitt, Geoffrey. (1998). "GS data processing methodology: from theory to applications," *GPS for Geodesy*, p. 231–270, Springer-Verlag, Berlin (1998).

Chandra A.M. (2005). "Higher Surveying" New Age International Publishers. ISBN: 8122416284.

Datta, D, Alam, J, Umeda, K, Hayashi, M, and Hironaka "Flood simulation a case study in the lower Mekong river basin" *Hydrol. Processing* 21, 1223–1237.

Eldho, T.I. Jha, and Singh, A.K. (2006). "Intergrated watershed modelling using a finite element method and GIS approach". *Intl. J. River Basin Management* 4(1).

Fell, P.J. (1980). Geodetic Positioning Using GPS, The Ohio State University Report No. 299, Columbus, Ohio, USA.

Hoffman-Wellenhof, B, et. al. (1997). *GPS Theory and Practice*, Springer Wien, New York.

Hugentobler, U., Schaer, S., and Fridez, P. (2001). "Bernese GPS Software Version 4.2 Documentation", Astronomical Institute University of Bern, Switzerland.

Kanetkar, T.P., and Kulkarni, S. V. "Surveying and Levelling" Pune Vidyarthi Griha Prakashan.

Kulkarni, M.N. (1999). "The GPS data processing for seismic hazard assessment". *Indian Surveyor.*, 52(1), 28–29.

Kulkarni, M.N. (2000). Edited, The GPS and Its Applications, Training Volume, GPS Training Course, Civil Engg. Dept., I.I.T. Bombay, 9–19 May.

Lhomme, J., Bourvier, C., and Jean Louis, P. (2004). "Applying a GIS based geo-morphological routing model in urban catchments". *J.Hydrol.*, 299, 203–216.

Rao, B.V., and Rao, E.P. (1988). "Surface Runoff modelling of small watershed" Proc. International Seminar on Hydrology of Extremes, Roorkee,Uttharakhand india, 145–155.

Rao, E.P., and Rao, B.V. (1997). "Surface runoff modelling of a watershed with land use from remotely sensed data." Workshop on Remote Sensing and GIS Application in Water Resources September CBIP, Banglore, India.

Reddy, K.V., Eldho, T.I., Rao, E.P., and Hengade, N (2007). "A kinematic wave based distributed watershed model using FEM.GIS and remotely sensed data". *Hydrol. Processes*, 21, 2765–2777.

Roux, H, and Dartus, D (2006). "Use of parametes optimization to estimate a flood wave: potential applications to remotes sensing of rivers". *J Hydrol* 328, 258–266.

Shahapure, S.S., Eldho, T.I., Rao, E.P. (2010). "Coastal urban flood simulation using FEM,GIS and remote sensing"*Water Resour Manage.*, 24, 3615–3640.

Shahapure S.S, Eldho T.J., and Rao E.P. (2011). "Flood simulation in an urban catchment of Navi Mumbai city with detention pond and Tidal effect watershed". *Port Coastal Ocean Eng., ASCE*, 137(6), 286–299.

8 Conclusions

Parikshit N. Mahalle

8.1 SUMMARY

This section concludes the book and proposes the future outlook that can be explored further to initiate new research avenues. This book intends to cover big data analytics, its challenges, and the main building block of data science, particularly in the context of civil engineering and allied areas. Today, the data is everywhere and it has become an integral part of our day-to-day lives. The amount of digital data that has been created during digital communication is growing exponentially. The data can be structured or unstructured and is being generated with a high velocity. Hence, the data is now big data and the traditional methods for modeling and providing optimal solutions to complex problems require a high amount of computing resources. Hence, the alternative effective solution is the use of data science–based solutions for all fields, including civil engineering. Data science is an interdisciplinary field that emphasizes data handling methods, processing techniques, and scientific algorithms to gain insights and knowledge from the huge amount of mixed data of civil engineering.

This book is especially for novice designers to give them a feel and understanding of how effective data science can be carried out in civil engineering. The tremendous growth of data and the need to envision hidden information from them makes it necessary to develop tools that will unleash the data. This book will help the civil engineer and other stakeholders to understand the requirement of data science in civil engineering and how to use it in the important field of civil engineering. This book discusses civil engineering problems like air, water, and land pollution; climate crisis; transportation infrastructures; traffic and travel modes; mobility services; etc. This book has two sections. The first section dealt with the basics of data science, essential mathematics for data science to be used in civil engineering. The next section dealt with applications of data science in fields like structural engineering, environmental engineering, construction management, and transportation, etc.

8.2 BUSINESS INTELLIGENCE

Business intelligence deals with analyzing the data that will have an impact in business decisions. BI is concerned with source data collection, transformation, and then making some critical predictions. Smart devices and new technologies are generating massive amounts of data. Data is an integral part of Industry 4.0, which has

DOI: 10.1201/9781003316671-8

key elements as artificial intelligence and big data. Using AI-empowered algorithms, civil engineering is overcoming hindrances and improving efficiency and productivity. Cost-effective and rapid project development in civil engineering is taking deep investments in the civil sector. The important aspects of civil engineering like quality management, risk control, maintenance, optimizations in design, etc. are well considered using data science. For an instance, the civil stakeholders can use data science concepts to understand the progress of construction projects. AI-powered recommendation systems can help civil engineers to suggest relevant information like selection of welded connections, superior architectures, etc. Artificial intelligence provides useful measures to assess risk by analyzing construction information and, in turn, this also helps the construction firms to prepare appropriate contingency plans.

8.3   RESEARCH OPENINGS AND FUTURE OUTLOOK

Consider an example of a real-world and complex flyover bridge. We have seen several news stories or posts of collapsed flyovers and their debris crushing the vehicles under it. The traditional way to construct a flyover bridge is a physical-based approach that has practical shortcomings. This includes the quality of construction, health monitoring, etc. However, with the advancement in information and sensing technology (wireless sensor network), it is possible to monitor and read a large number of vital parameters of complex flyover bridges continuously by implementing AI-powered wireless network systems. Using this system, we can read, store, and analyze historical data of flyovers. The analyses of historical data help the construction and maintenance team to understand the quality of a flyover bridge, the associated risk, and other important parameters to avoid a flyover bridge collapse. Due to increasing data size and its increasing complexity, data-intensive research is going to be a major research area for the next few decades. Data intensity comes with the pool of challenges and design issues especially in the context of research and development, as data science and data storytelling do not go alone. It should be linked to some use cases so that an appropriate data set can be explored for the purpose of storytelling and analytics. The prominent research areas that can be explored further are listed below:

- Explainable artificial intelligence for driving use cases
- Predictive analytics for preventive approach
- Data science for affective computing
- Storytelling and visualization
- Gamification for civil engineering
- Data exploration
- Prescriptive analytics

In addition to this, there are other allied areas like data acquisition, data preparation, data migration, data extraction, loading and transformation, etc. that also need to be explored further. As a takeaway from the book, one must be motivated to look into futuristic opportunities in this world of data science, particularly for civil engineering.

Index

Printed in the United States
by Baker & Taylor Publisher Services